Developmental Work Research

Developmental Work Research

A tool for enabling collective agricultural innovation

Mutizwa Mukute

Wageningen Academic
Publishers

EAN: 9789086862733
e-EAN: 9789086868193
ISBN: 978-90-8686-273-3
e-ISBN: 978-90-8686-819-3
DOI: 10.3921/978-90-8686-819-3

First published, 2015

Wageningen Academic Publishers
P.O. Box 220
6700 AE Wageningen
the Netherlands
www.WageningenAcademic.com
copyright@WageningenAcademic.com

Table of contents

Chapter 2. The thinking behind Developmental Work Research 47

Chapter 3. Illuminating the DWR methodology 87

Acronyms and abbreviations

AIS	Agricultural Innovation Systems
AISA	Agricultural Innovation Systems in Africa
AKIS	Agricultural Knowledge and Information Systems
BoP	Base of the Pyramid
BRICS	Brazil, India, China and South Africa
CBNRM	Community Based Natural Resources Management
CELUCT	Chikukwa Ecological Land Use Trust
CGIAR	Consultative Group of International Agricultural Research
CHAT	Cultural Historical Activity Theory
CL	Change Laboratory
CRADLE	Centre for Research on Activity, Development and Learning
DEVCO	European Commission's Directorate General, Development and Cooperation
DWR	Developmental Work Research
ELRC	Environmental Learning Research Centre
FAO	United Nations Food and Agriculture Organisation
FARA	Forum for Agricultural Research in Africa
GFAR	Global Forum for Agricultural Research
IAR4D	Integrated Agricultural Research for Development
ICRA	International Centre for development-oriented Research in Agriculture
INSARD	Including Smallholder Farmers in Agricultural Research for Development
IPs	Innovation Platforms
ISCAR	International Society for Cultural and Activity Research
JOLISAA	Joint Learning in Innovation Systems in African Agriculture
MADF	Machobane Agricultural Development Foundation
MFS	Machobane Farming System
MKO	More Knowledgeable Other
NARS	National Agricultural Research Systems
NEPAD	New Partnership for Africa's Development
NGOs	Non-Governmental Organisations
NRM	Natural Resources Management
NULC	Nyahode Union Learning Centre
PELUM	Participatory Ecological Land Use Management Association

PLA	Participatory Learning and Action
PRA	Participatory Rural Appraisal
PROLINNOVA	Promotion of Local Innovation
PTD	Participatory Technology Development
RSDA	Rural Self Development Association
SADC	Southern African Development Community
SAQA	South African Qualifications Authority
SARDC	Southern African Research and Documentation Centre
SCOPE	Schools and Colleges Permaculture Programme
SSA CP	Sub-Saharan Africa Challenge Programme
SSA	Sub-Saharan Africa
TSURO	Towards Sustainable Use of Resources Trust
UNDESD	United Nations Decade of Education for Sustainable Development
UNIDO	United Nations Industrial Development Organisation
ZPD	Zone of Proximal Development

Acknowledgements

I would like to acknowledge many groups of people and organisations for their respective roles in inspiring and enabling me to write this book. The groups of people who have inspired me include:

The small scale farmers who have received and passed on agricultural knowledge from one generation to another; and especially the farmer innovators, such as Machobane in Lesotho who have sought creative solutions to the complex agricultural issues.

The promoters of farmer innovations and innovators that include: the Organic Conservation Agriculture project in Zimbabwe, Phiri Award for Farm and Food Innovators, Machobane Agricultural Development Foundation (MADF), Promotion of Local Innovation (PROLINOVA), Participatory Ecological Land Use Management Association (PELUM), Honey Bee, the Eastern and Southern African Farmers' Forum (ESAFF), Forum for Agricultural Research and Development (FARA) and Global for Agricultural Research and Development (GFAR).

Cultural Historical Activity Theory (CHAT) scholars whom I met at the International Society for Cultural and Activity Research (ISCAR) symposium that was held in Rome in 2011 who include: Professor Harry Daniels, Professor Yrjö Engeström, Dr Roberta Hill and Dr Annalisa Sannino.

Leading scholars on Cultural Historical Activity Theory (CHAT) who critiqued my drafts and provided relevant literature, Professor Emeritus Jaakko Virkkunen, Professor Yrjö Engeström and Dr Annalisa Sannino of the Centre for Research on Activity, Development and Learning (CRADLE) at the University of Helsinki.

Finally, Rhodes University scholars, especially Professors Heila Lotz-Sisitka and Rob O'Donoghue of the Environmental Learning Research Centre (ELRC) for insisting that I pitch the book at the right level. My Rhodes University college mates provided constructive feedback on my drafts and also availed their study reports for use in the book. These include Dr Lausanne Olvitt, Mr Chris Masara, Dr David Lindley, Dr Dick Kachilonda, Mr Tichaona Pesanayi and Dr Million Belay.

Chapter 1. Introduction

1.1 Background

There is no society that can survive without agriculture. Agriculture is deeply rooted in human traditions, cultures and histories. In southern Africa, where this book was produced, farmers are the mainstay of most economies. They spend most of their productive time tending the soil, watering the plants, harvesting produce and looking after their livestock. Sometimes they select seed for the next season; others try new ways of practising agriculture. They do all this to produce enough food for consumption and surplus for income generation. They often complement their agricultural activities with looking after and utilising their natural resources such as watershed, forests and woodlands, rivers and dams.

Over the years these farmers across the world have developed much that is deeply embedded in their practice. However, the world around us is changing. Rains are coming later and in smaller quantities, soils are being depleted, and rivers are running dry sooner than in the past. The population that farmers must feed is growing fast while arable and grazing land is diminishing. Seeds are being genetically transformed and farmers are losing some of the control that they used to have over the agriculture value chain. Meanwhile youths are losing interest in farming, preferring professions that are 'cool'. These and other issues create the need for ongoing learning and adaptation amongst farmers and other agricultural stakeholders. Farmers, like many others in society are also constantly innovating, and where there is a need to improve food security in the face of new risks, such innovations need to be expanded, shared more widely with others.

To support farmers, the agricultural sector has traditionally trained extension workers who have developed different approaches to working with farmers. However, farmers need to work with many other stakeholders as farming systems change, and challenges and solutions emerge. This resource book introduces a new approach to supporting agricultural innovation that recognises the cultural histories and experiences of the farmers. It is rooted in the realities facing farmers today and in the possibilities for shared learning, innovation, action and development with fellow agricultural stakeholders. This method is called 'Developmental Work Research' (DWR). It is a theoretically informed approach that has been tested in many different countries, and has potential to strengthen agricultural innovation platforms in Africa as is shown by the stories that permeate the text. Textbox 1.1 shares the story of a Zimbabwe small scale farmer who innovated in response to the effects of cyclone Eline.

Textbox 1.1. William Gezana's story of innovation.

When cyclone Eline hit Chimanimani district in 2000 it destroyed homes, infrastructure and massively eroded fields, leaving them barely arable. Many affected villagers were resettled through government assistance but I decided to stay put. I knew that I could draw on my farming experience to heal the land and make it productive again. At the same time I drew on the knowledge and skills that I had acquired from working with government agricultural extension workers and NGOs. These include Chikukwa Ecological Land Use Trust (CELUCT), Towards Sustainable Use of Resources Trust (TSURO) and Nyahode Union Learning Centre (NULC). I also learnt from fellow farmers. My primary task was to find ways of producing enough food for my family after the cyclone had removed top productive soil. Using simple tools to work the land, I was able to do the following:

- design irrigation channels to carry water from the river that was at the bottom of my field;
- reclaim degraded land by trapping soil using bands of stones and vertiver grass; making compost from leaves, grass and animal manure; resting the land and allowing soil to form through natural regeneration processes;
- utilise parts of the land where top soil had remained or accumulated and introduce crop rotation to overcome diseases that arose from the mono-cropping of tomatoes;
- learn about and implement fruit-tree grafting and establish an orchard; and
- propagate indigenous trees and mobilise fellow villagers to manage the watershed and reduce bare ground, soil and water loss; this was accompanied by the establishment of small scale fish farming in the river.

Within a decade I was able to restore and increase the productive potential of my land, recharge the water table and have water available for irrigation. My family produces surplus food for sale. About 40 households in the village have learnt from me and with me and more importantly many of them have adopted the innovations. However, some villagers used to question the wisdom of my efforts and only took me seriously when they saw the good results. Now I am involved in a project that is looking at how to overcome the witch weed problem.

1.2 Innovation as a concept

People who are involved in agriculture and Natural Resources Management (NRM) have to face the questions of food and nutrition security, maintaining and improving the quality of productive land and water bodies and managing the agricultural value chain. Most of the answers to the questions must be sought continuously, through innovation. Innovation is especially important because of the need for new solutions in the face of rapid population growth, high food prices, climate change and diminishing oil for powering agriculture. This book is inspired by the need to make innovation in agriculture and NRM more widespread by explaining how groups of people in a shared environment and common interests can jointly innovate.

Since innovation is a central concept in this book, it is necessary to define it as part of setting the scene. *Innovation* is a process of bringing about novel social-institutional and scientific-technological changes in cultures, organisations and human practices. It is a complex, multifaceted and dynamic process involving multiple and changing stakeholders, interacting intimately with their environment, evolving over long periods and unfolding in directions impossible to devise from the start (Triomphe *et al.*, 2014: p. 54). It enables people to negotiate the future based on the recognition that there are limits to planning. Innovation is based on individual and collective human agency; and individual and collective learning across sectors, disciplines and levels. It embraces uncertainties inherent in the future, while seizing the opportunities that it offers by going beyond the known solutions. It is aimed at changing the production processes or introduction of new products, whose details cannot be forecasted at the beginning of the innovation process. The intended outcome of on-going innovation is iterative human progress, which produces innovative solutions to potential and prevailing problems, including those arising from human advancement. Such risk-reducing innovation could be achieved through the production of improved and safe technologies in agriculture and other economic practices (Nowotny, 2006).

In the context of DWR, the core of an innovation is a tool or a method that enables actors to resolve a contradiction that historically evolved in their practice or activity or group. For example, Gezana, the innovative farmer discussed in Textbox 1.1 developed methods to resolve a contradiction between his family's need to cultivate land and produce enough food for the family on one hand and the eroded productive capacity of its fields on the other. Put simply, innovation is a process by which new ideas are transformed into practice (Hall, 2010) through a process that DWR refers to as ascending from the abstract to the concrete (Engeström *et al.*, 2014).

The Forum for Agricultural Research in Africa (FARA) defines innovation as 'the activities and processes associated with the generation, product distribution, adaptation and use of new technical and institutional/organisational knowledge' (Adekunle *et al.*, 2014: p. 7). Similarly Chema *et al.* (2002), define agricultural innovations as activities and processes associated with the generation, production, distribution, adaptation, adoption and widespread use of new technical, institutional, organisational or managerial knowledge in the agricultural sector. Agricultural innovation is the process whereby individuals or organisations bring new products, processes and forms of organisation into social and economic use to increase effectiveness, competitiveness, and resilience to shocks or environmental sustainability, thereby contributing to food and nutrition security, economic development and sustainable natural resources management (FAO, 2012). *Agricultural innovations* are activities, processes and products associated with the generation, production, distribution, adaptation, adoption and widespread use of new technical, institutional, organisational or managerial knowledge in the agricultural sector (Chema *et al.*, 2002; Pretty, 2003).

1.2.1 Innovation systems and partnerships

An *innovation system* is a network of actors and organisations that collectively innovate and convert the innovations into productive use (DEVCO, 2014). The World Bank (2008) sees an innovation system as a network of organisations, enterprises and individuals focused on bringing new productions, new processes and new forms of organisation into economic use, together with the institutions and policies that affect their behaviour and performance. Agriculture and NRM are complex practices that involve multiple actors along the value chain. For example, agriculture requires working across disciplines covering crops, livestock, soil, rainfall, agro-processing and agro-chemical development; and NRM includes geography, ecology, wildlife management, fisheries and water resources management. The complexity of these practices makes dialogue and collaboration across disciplines essential thus making innovation partnerships necessary for the production of effective solutions. Complexity is present when a system or practice has many elements that interact with each other and produce outcomes that are difficult to predict. Bringing groups of people to develop complex systems and practices requires enabling methodologies such as Developmental Work Research (DWR), which is what this book is about.

1.2.2 Agricultural innovation system

An agricultural innovation system comprises many interacting stakeholders using their collective capacities and relationships to tackle a complex agriculture-related challenge to produce and utilise a solution. This should result in a developmental and qualitative change in system and/or practice. An agricultural innovation systems approach is gaining traction because of the following reasons, among others:

- The realisation that scientists are not the only innovators and that other actors such are farmers play an important role in innovation, often innovating with alternative modes of practice. Multiple knowledge bases enable successful agricultural innovation, going beyond the dichotomy of scientists and farmer knowledge to include a more diverse set of actors and organisations;
- Changing paradigms in development practice, where participation, diversity and self-reflection are becoming the expected modes of professional behaviour; and
- The broadening agricultural research agenda to include poverty reduction and environmental sustainability, which are complex issues (Hall *et al.*, 2004: p. 7).

In Finland the collaboration between farms and providers of farm materials and equipment as well as the collaboration between farmers and their industrial clients has turned out to be important sources of innovation. This kind of innovation tends to get stimulated in instances where a value chain development approach is used – from 'farm to fork' in the context of food products in agribusiness. Wiggins & Roepstorff (2011: p. 28) define agribusiness as 'a broad concept that covers input suppliers, agro-processors, traders, exporters and retailers. Agribusiness provides inputs to farmers and connects them to consumers through financing, handling, processing, storage, transportation, marketing, and distribution of agro-products.' Wohlmuth (2011) points out that value chain development impulses can be driven from below by local producers, intermediaries and processors and for this to happen there ought to be linkages and coordination with the private sector. Experiences in Finland suggest that providing farmers with new toolkits to plan their production and optimise selection of crops is an important factor in stimulating collaborative learning between industry and farmers (Von Hippel, 2001).

1.2.3 Themes or domains covered by agricultural innovation

The European Initiative for Agricultural Research for Development (EIARD) identifies several domains of Agricultural Research for Development, which should also be considered in agricultural innovation and these are: (1) crop production; (2) animal husbandry; (3) agro-forestry; (4) fisheries and aquaculture; (5) agribusiness and related enterprises; (6) animal- and human-health related issues; (7) sustainable management of the natural resources on which farming depends; (8) socio-cultural and bio-diverse landscapes; and (9) food systems and ecologies. Mukute and Marange (2011), in their mapping study for Including Smallholder Farmers in Agricultural Research for Development (INSARD), added two more: (10) *land tenure and rights*; and (11) *mechanical engineering*. These domains are important for defining the scope of activities that can take place in agriculture.

1.3 Purpose and objectives of this book

This book seeks to explain DWR and how it can be used to strengthen multistakeholder learning, innovation and action in agriculture and NRM. It particular it discusses the background to, process and results of using DWR with a view to supporting:

- The bridging of top-down and bottom up learning, innovation and development approaches in the context of dualisms that stifle human advancement.
- Multistakeholder, transdisciplinary and transformation learning, innovation and action in agriculture and natural resources management context, which are important for most of Africa's agrarian societies and economies.
- Cognitive justice by enabling plural ways of knowing to be acknowledge and tapped into for the development of knew knowledge and innovations.
- The achievement of academic interests that are based on the rigour of the methodology, which is in turn inspired by its philosophical and theoretical foundation. It is underpinned by dialectics and an ontology of developing systems and was developed in the context of Cultural Historical Activity Theory (CHAT).

Its intended practical purpose is therefore to contribute to the continuous and comprehensive improvement of agricultural and natural resources management in the context of sustainable development whose triple objectives are economic viability, social justice and ecological sustainability. One of its academic interests is to enhance the relevance of higher education and agricultural research and innovation to society, with a particular interest in the African agrarian communities.

Dialectical thinking links and goes beyond top-down and bottom-up approaches, theory and practice, the abstract and the concrete, and the individual and society. DWR grew out of, and is most useful in improving productive work such as agriculture and NRM. This means that it is especially useful in enabling workplace learning and transformation. Its other strength lies in providing a systemic and historical view based on a theoretical foundation that helps its users to overcome dualisms (Engeström, 1987, 1999). Such dualisms were identified as obstacles to agricultural research and extension where top-down approaches have been pitted against bottom-up approaches because of lack of a bridging theory (Leeuwis, 2004). DWR works with principles and tools that enable collective learning, innovation and agency development among people with different histories, identities, interests, resources and knowledge.

The specific objectives of the resource book are therefore to:
- link current agricultural and NRM innovation discourse to DWR;
- encourage the use of DWR to support workplace research and rural innovation systems;
- explain the theoretical concepts and principles that underpin DWR to help current and future researchers who are interested in workplace learning, knowledge generation and innovation;
- provide guidelines on how to stimulate collective learning, innovation and agency development using DWR and its associated method called Change Laboratory (CL) in the context of Integrated Agricultural Research for Development (AR4D), Agricultural Innovation Systems (AIS), People Centred Learning and Innovation, and Innovation Platforms; and
- share lessons and insights from the implementation of DWR in agricultural and NRM contexts.

1.4 Audience

The book is primarily intended for people who facilitate multistakeholder learning and innovation processes in agriculture and NRM. These include students, researchers and educators who are interested in a theory-informed interventionist research methodology that produces practical change while at the same time generating new knowledge. More specifically, the book aims at the following audiences:
- Masters and PhD students interested in theorised Action Research and Participatory Action Research, especially in the field of agriculture, natural resources management and education;
- agricultural research and development researchers who are interested in collaborative partnerships with other stakeholders, such as farmers and the private sector;
- agricultural extension workers and development facilitators who want to broker knowledge sharing and development, and innovation processes;
- universities who want to strengthen their community outreach programmes in such a way that they generate academic value for themselves and practical value for the communities that they choose to work with; and

- agricultural research institutions who are looking for more robust ways of making agricultural innovation platforms work better.

In general, it is intended for people who want to understand and improve workplace learning to improve practices, products and processes.

1.5 Justification

Multistakeholder agricultural innovation has become imperative in the face of food and nutrition insecurity, climate change and population growth; and the need for development that is sustainable. The importance of multi-stakeholder innovation processes is acknowledged by the United Nations Food and Agriculture Organisation (FAO), which views the roles of the main innovation stakeholders as follows:

> *Farmers need to innovate in the systems they use; governments need to innovate in the specific policies they implement to support family farming; producers' organisations need to innovate to respond better to the needs of family farmers; and research and extension workers need to innovate by shifting from research driven process predominantly based on technology transfer to an approach that enables and rewards innovation by family farmers themselves. Additionally, in all its forms, innovation needs to be inclusive, involving family farmers in the generation, sharing and use of knowledge so that they have ownership of the process, taking on board both the benefits and the risks, and making sure that it truly responds to local contexts.*
>
> *(FAO, 2014: p. viii)*

Writing this book was inspired by the knowledge that there are currently no books that link DWR to agro-ecological innovation systems and processes. And yet the methodology has potential to strengthen these innovation systems and processes. I was specifically motivated by my rural roots and:

- my professional experience in NRM and agriculture;
- the need to share a theory and methodology that transcends top-down and bottom-up learning and development approaches; the methodology also enhances collective learning and innovation; individual, relational and collective agency; and
- the potential value of disseminating DWR-inspired research insights in agriculture and NRM.

1.4.1 Personal experience in NRM and agriculture

When I worked as a national parks manager and extension work in Zimbabwe, I experienced the tension between wildlife protection and community access to benefits of biodiversity conservation. My approach was primarily top-down. Later, as an environmental researcher-writer of the Southern African Research and Documentation Centre (SARDC), I learnt that the same clashes were taking place across the Southern Africa Development Community (SADC) region. At the same time I came to know about Community Based Natural Resources Management (CBNRM) initiatives that were being made to address the clashes.

Later, as the college coordinator, and director of Participatory Ecological Land Use Management Association (PELUM) I worked with inclusive and bottom-up approaches. My approach was shaped by Participatory Rural Appraisal (PRA), Participatory Learning and Action (PLA), people-centred development, pedagogy of the poor and action research. I coordinated the development and implementation of an agro-ecology and community development curriculum by multiple actors. These were lecturers from two universities in Zimbabwe; trainers from two government departments; and development facilitators from about 10 NGOs. We used both top-down and bottom-up training approaches. I used action research to evaluate coherence in the curriculum design and implementation for my Masters half-thesis. I faced the challenge of finding a theoretical framework that could sufficiently link the predominantly top-down training that was being implemented through universities and bottom-up training that NGOs used. I met Professor Anil Gupta from India who shared his experience on documenting farmer innovations and used his insights to facilitate the documentation of about 20 farmer innovators in southern Africa. Two of them were awarded honorary doctorates for their respective innovative work: Dr Chinkhuntha from Malawi for his work on transforming swamps and gullies into productive land (Chima & Chinkhuntha, 2002), and Dr Machobane for developing sustainable agricultural practice over 13 years (Machobane & Berold, 2003). These experiences convinced me that farmers do contribute to agricultural innovation and should be treated as partners in agro-ecological innovation.

During the same period, I became a member of the NGO Committee of the Consultative Group of International Agricultural Research (CGIAR), which faced challenges of cooperation between agricultural researchers and farmers. Farmers were largely excluded from agenda setting. The main challenge appeared to be the same: how to reconcile top-down and bottom-up approaches to agricultural research and development. One of the major global responses to this development was the formation of the Global Forum for Agricultural Research (GFAR) – a multistakeholder forum that seeks to transform agricultural research, education, extension and enterprise for development. It comprises a wide range of stakeholders including: international agricultural research centres, farmer organisations, NGOs, private sector, donors and investors, advisory services, education services and youth organisations. I also participated in the initial stages of the formation of Promotion of Local Innovation (PROLINNOVA) – a multistakeholder international network that is at the forefront of promoting local innovation and farmer-lead participatory research and development processes.

It was during and after my PhD studies that I came across DWR and realised its merits in dealing with the learning, innovation and transformation challenges that I had encountered during my professional work (Mukute 2010a, 2013). I felt convinced that sharing the methodology has the potential to improve agricultural and NRM learning, innovation and transformation in the workplace. The need for supporting farmers to innovate on their own and with others has also been highlighted elsewhere. For example, a 2014 CGIAR System Research Programme workshop notes that:

> *Communities whose livelihoods come at least partly from agriculture and natural resources management are in the midst of a rapidly changing world ... When actors at different levels of agricultural research and development – ranging from resource-poor women farmers to international agricultural research and development policy makers – can better interact, they bring different perspectives and insights into the process of understanding the dynamics of the system ... This makes them better able to react quickly, flexibly and creatively to shocks, challenges and opportunities ... Focusing on resource-poor and vulnerable people's capacity to innovate – that is, on inclusive innovation – can change the power balance, so that these people can recognise more possibilities, unfold their innate creativity, and more confidently tackle newly emerging problems and opportunities for a more equitable world.*
>
> (*Leeuwis* et al.*, 2014: p. 6*)

Navi Radjou's writing on the Jugaad – an innovative approach that initially inspired the Industrial Revolution in the West and that is driving economic growth among the BRICS (Brazil, Russia, India, China and South Africa) nations – has also inspired me. His work on innovation is based on six principles, which are:

- seeking opportunity in adversity by treating harsh constraints as invitations to innovation and the creation of value;
- doing more with less by being resourceful and frugal in the face of scarcity;

- thinking and acting flexibly by constantly questioning conventional wisdom, working outside the box and beyond linear thinking, responding to unexpected changes and transforming existing products, services and processes;
- keeping it simple by developing a good enough solution that gets the job done.
- including the margin by creating radically affordable solutions that include the Base of the Pyramid (BOP) – the low income and non-traditional customers, not as passive consumers by as active co-creators; and
- following the heart by employing intuition, empathy and passion and not relying on cold facts and figures alone; the main driving force of Jugaad innovations is not the money but a desire to make a difference in the community (Radjou *et al.*, 2012).

1.4.2 The need for sharing a theory and methodology that transcends top-down and bottom-up approaches

Most of past development approaches were based on farmers and natural resource users being told what to do and how to do it through top-down development approaches. The response was the development of bottom-up approaches, which sought to 'put people first' and do development from the bottom (Chambers, 1997). But then there are also challenges associated with this dualism, which assumes that the two are mutually exclusive of each other. DWR overcomes this dualism as it was formed in the context of Cultural Historical Activity Theory (CHAT) – a theory that is underpinned by dialectics. Experience in agricultural research and extension suggests that both top-down and bottom-up make contributions to the development of agricultural practices (Mukute, 2010b, 2013). However, there have been challenges associated with how to work with both approaches in a coherent way. CHAT is a learning, development and agency development theory, which encompasses intergenerational knowledge transmission, learning from those who know more and collective generation of knowledge and innovations.

Making DWR accessible to a wide range of learning, innovation and development facilitators has the potential to create a critical mass of people who can use the methodology to mobilise the knowledge and agency of relevant multistakeholders to innovate for a better world.

1.4.3 The need to utilise DWR-inspired research insights in societal development

The potential of CHAT-informed DWR to support agriculture and NRM multi-stakeholder learning, innovation and agency development processes has been illustrated by studies that have been conducted in Finland, New Zealand, Ethiopia, Lesotho, Malawi, Mozambique, South Africa and Zimbabwe. The studies suggest that DWR can contribute to the improvement of agricultural and NRM practices. Some of the studies that have used DWR in the rural development sector are outlined in Textbox 1.2.

Textbox 1.2. Outline of studies that used DWR.

Agricultural sector
- learning challenges in vegetable farming in Finland (Seppänen, 2004);
- workplace learning in the New Zealand apple industry network (Hill *et al.*, 2007);
- exploring and expanding sustainable agriculture practices in Lesotho, South African and Zimbabwe (Mukute, 2010a,b; Mukute & Lotz-Sisitka, 2012);
- collaborative innovation and transformative agency in (agricultural) primary production in Finland (Vänninen, 2012);
- collective learning and agency development in newly established organic farmer associations in Zimbabwe (Mukute, 2012, 2013); and
- expansive learning for resilience in agricultural and water sustainability practices in South Africa and Zimbabwe (Pesanayi, 2014).

NRM sector
- ethics-oriented education in environmental education workplaces: an activity theory approach, South Africa (Olvitt, 2010);
- social learning processes and nature-culture relations in commercial beekeeping in South Africa and Zimbabwe (Masara, 2010);
- participatory learning and change in the context of bio-cultural diversity and resilience in Ethiopia (Belay, 2012);
- investigating and expanding learning in co-management of fisheries resources to inform extension and training in Malawi (Kachilonda, 2014); and
- expansive learning processes and strengthening of organisational learning for improved wetland management in a plantation forestry company in South Africa (Lindley & Lotz-Sisitka, 2014).

Energy sector
- learning challenges in biogas production for sustainability in Finland (Pereira-Querol, 2011); and
- developmental possibilities of environmental activities focusing on on-farm biogas production in Finland (Pereira-Querol *et al.*, 2014).

Vänninen's (2012) study illustrates how DWR has and can be used to inspire agricultural innovation. The study is based on overcoming whitefly pest problems in tomato and cucumber production in western Finland. The problem arose when farmers in the region used greenhouses (which trap heat) to ensure the production of horticultural crops throughout the year in a country that becomes too cold for agricultural production in winter. But the greenhouse also resulted in a rapid growth of the pest populations based on warmer temperatures and food availability. The pests spread rapidly because the greenhouses were built close to each other. The initial strategies for addressing the problematic situation were compartmentalised and ineffective. This prompted the pest management researcher to use the DWR approach. Consequently the farmers, advisors, agricultural researchers and a representative of buyers of the packinghouse held a CL workshop. Below is a summary of what happened through the DWR-informed workshops:

- The analysis showed that although growers, researchers and advisors were working with the same object, the whitefly, they conceptualised and approached it in different ways, which made collaboration difficult. They were working in isolation, without sharing information and coordinating their actions. Growers did not, for instance, inform their neighbours when they cleaned their greenhouses (which made the whiteflies move) so that the neighbours could prevent the whiteflies from entering their greenhouses.
- During the first three CL sessions a model was constructed depicting the reproduction and spreading of the pests in the area and a hypothesis was created about the historical contradictions behind the growing of the pest problem: one between actual information used for taking decisions on controlling the whitefly (tool) and the unpredictable and explosive multiplication of the pest (the object) and the other between division of labour in which each grower took decisions only within own farm, and the characteristic of the pest that did not 'bother themselves of the borders'. The participants developed a solution based on four elements:
    1. Systematic monitoring and information sharing about pests among growers.
    2. A tool for sharing information about the pests.
    3. Collaboration between growers, advisors and researchers in producing the necessary knowledge for making pest management decisions.
    4. Regular meetings with representatives of growers, advisors, and researchers to follow up and analyse the situation and plan the necessary further actions.
- The participants solved the problem of different conceptualisations and approaches by starting to use a monitoring system that had been created for research purposes (Vänninen, 2012; Virkkunen & Newnham, 2013: p. 233).

Textbox 1.3 gives an example of how DWR was used in a South African NRM setting (Lindley & Lotz-Sisitka, 2014) to stimulate collective learning and innovation.

Textbox 1.3. DWR inspired collective learning and innovation in NRM.

DWR was used in developing innovative solutions to problematic situations that were being faced in the management of Mondi wetlands in South Africa. Wetlands are important for purifying water, regulating water flow, recharging groundwater, controlling floods and erosion and for providing livelihood opportunities for local communities. The opportunities include fishing, grazing, agriculture and tourism. But the wetlands were threatened by reduced water inflow from its catchment as a result of forestry plantations, which consume more water than the grasslands they replaced. Other major threats to the wetlands were poor burning practices, poorly managed livestock grazing, inappropriate location and construction of roads in wetlands, and inadequate control of alien plant species. The management of the wetland is a shared responsibility between foresters who manage plantations around the wetland, environment and conservation education officers who manage the wetlands and community engagement facilitators who work with communities that live around and derive livelihoods from the wetland. In spite of capacity building actions that were conducted over several years, the wetland continued to degrade.

A DWR-inspired interventionist study that was done identified and tackled two main problems that were being faced by the three stakeholder groups. These were the absence of a mechanism to enable effective learning of wetland management by the practitioners; and lack of mechanisms to overcome compartmentalised approach to wetland management. These challenges were overcome through CL sessions that involved representatives from the three stakeholder groups and resulted in the following improvements:

- strengthening the scope, depth, and sophistication of participant understanding of the situation;
- expanding the ways staff interact and work together;
- democratising decision making;
- improving social relations between staff through reducing the power differentials, and creating stronger relationships; and
- enhancing participant reflexivity through deeper understanding of the social structures and cultural systems, and changing them to support improved wetland and environmental practice of staff, and developing the organisational structures and processes to strengthen organisational learning and development.

Consequently, it catalysed organisational learning and development changes among the three stakeholders and resulted in improved wetland management practices in Mondi.

Source: Lindley & Lotz-Sisitka (2014).

Another interesting DWR-inspired innovation took place in the apple industry of New Zealand where the Centre for Research on Work, Education and Business Ltd (WEB Research) and the New Zealand Department of Labour led a research project. The research project was funded by government to address tensions in the country's apple industry by creating new forms of regulatory activity through collaboration between government and industry (Textbox 1.4).

Most of these studies confirm that:

- agricultural and NRM cognition is found in both the practitioner and the scientist;
- agricultural and NRM cognition is based on the interplay of local, situated knowledge and outside generic knowledge; and
- CHAT-informed DWR, which can effectively enable collective innovation in agriculture and NRM, is under-utilised.

Some of the studies also illustrate how DWR enables researchers and research participants to deal with complex agricultural development issues. These complexities include the:

- ecological, material and knowledge resources that farmers have;
- different and related motives of farmers and resource users, technology and input suppliers, extension workers, researchers, policy makers, food processors and consumers;
- different and sometimes competing agricultural practices such as traditional, conventional and sustainable agriculture that are promoted by different organisations and people;
- desire to ensure the continued productive capacity of the land while ensuring ecological viability and fair distribution of costs and benefits arising from agriculture; and
- need to keep adjusting to changes in the social and ecological changes such as human induced climate change (Mukute, 2010a).

Stoop & Hart (2005) suggest that close collaboration between farmers and researchers is necessary for improvements in agriculture. This happens effectively when institutional arrangements enable partnerships between stakeholders with dissimilar skills and operating at different levels. Similarly, Wals & Bawden (2000) note that dealing with complexity associated with ecological sustainability requires a fundamental transformation in the competences required by multi-functional extension workers. Many of the social ecological problems can only be solved through collective decision-making, innovation and action.

Textbox 1.4. DWR-inspired research in New Zealand's apple industry.

The researchers note:

> *When the experiment in the apple industry started in 2004, the interests of the apple growers and the government were in conflict. To survive, growers need their apples to be picked, and rely on just enough seasonal labour to accomplish this annual 'harvest miracle' from any source that they can find. With one of the lowest unemployment rates in the OECD, it is harder to source labour legally now and grower anxiety has induced illegal action that government cannot ignore. The research team believed that all parties were interested in finding a solution as an alternative to the slow collapse of the industry.* (p. 361)

Apple growers, labour contractors, pack house operators, quality controllers, horticultural consultants and government officials participated in the DWR journey. The initial focus of the methodology was on immigration regulations, which affected labour availability, which in turn was important for the survival of the apple industry. The subsequent focus was on quality and the long-term sustainability of the New Zealand apple industry.

The DWR led to a deepened, shared understanding of the systemic contradictions that the apple industry needed to address, which included the fact that:

> *The industry will not survive if it remains reliant on illegal labour yet domestic workers alone are insufficient to supply current harvest labour needs; that apple harvesting is a skilled job that is best done by those trained and/or experienced, yet current policies favour the use of unskilled, untrained and inexperienced workers ...; that the industry needs to assume some responsibility in mitigating the risks involved in using temporary migrant labour despite contractors, sub-contractors and growers that cooperate with regulatory agencies often find themselves commercially disadvantaged; and finally, that the often highly prescriptive regulatory and customer demands from beyond New Zealand's border need to be recognised by New Zealand policy makers and implementers.* (p. 370)

The subsequent innovation process focused on five areas:
- A training strategy;
- Improving the quality of the product based on careful selection of apples as quality was a key export consideration;
- Research and development around production and labour practices, beyond apple varieties;
- Contractor-grower relationships; and
- Labour sourcing, deployment and retention strategy which had a bearing on immigration policy.

Source: Hill *et al.* (2007).

1.6 Key features of DWR

The DWR methodology has at least five features that distinguish it, and these are:

- a theoretically grounded unit of analysis called an activity system;
- an explicit theory of future oriented learning and development, based on contradictions as a potential source of learning, change and development of activities;
- a theory of innovation and knowledge formation, which is based on recognizing the essential difference between two forms of knowledge formation and generalisation: empirical-analytical and theoretical-explanatory and the productive value of the latter;
- a methodology for formative intervention, which is based on the general principle of ascending from the abstract to the concrete; and
- a theory and method of supporting practitioners' transformative agency.

1.6.1 Forms of learning supported by DWR

DWR is an interventionist research and development methodology that enables vertical and horizontal learning, knowledge generation, innovation and agency development. Vertical learning is concerned with *development* across levels, from the rudimentary to the advanced forms of a practice. Horizontal learning on the other hand is concerned with *development* across borders or boundaries. It focuses on developing shared meanings, solutions and actions among interacting people of different backgrounds. Pimbert (2009: p. 22) notes that such democratic ways of knowing are required 'to bring together partial and incomplete perspectives of different actors faced with uncertainty, diversity and dynamic change.'

DWR was developed in the context of CHAT, which offers three interpretations of learning, which link and integrate top-down and bottom-up learning. These are:

- *Cultural mediation*: This form of learning is based on cultural artefacts and their instrumental use. Such an artefact can be conceptual or material. The learning set-up primarily consists of the More Knowledgeable Other (MKO), who knows how to use the artefact, a novice who does not know how to use the artefact, and the artefact itself. The novice learns in mutual interaction with the MKO to understand the meaning of the artefact and how to use it as an instrument to enable the understanding and/or transformation of a situation. The MKO mediates the novice's learning by directing his/her attention to factors that are critical for succeeding in the task. This is based on the instructor/MKO's analysis and understanding of the learner's needs – the novice/learner's Zone of Proximal Development (ZPD). Figure 1.1 shows how the MKO helps the novice

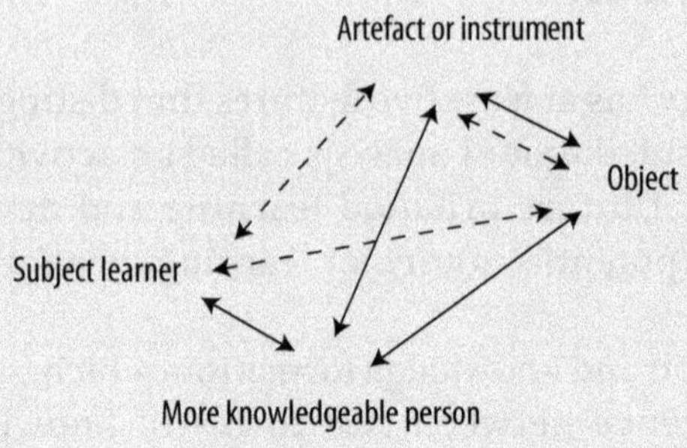

Figure 1.1. Cultural mediation of learning by a more knowledgeable other.

or learner (subject) to understand and use an artefact so that he/she can perform tasks that are pre-defined. The developmental aim of such mediated learning is achieved when the novice masters the concept or task (Tarulli & Cheyne, 2005).

- *Developing mature concepts*: The MKO helps the learner to convert everyday knowledge to mature concepts. In this regard Edwards (2005) notes that the MKO should help the learner to move from 'situated everyday understandings' to the understanding of associated scientific concepts that can be applied outside the contexts in which they were developed. Gramsci (1971) notes that such learning is concerned with transforming laypeople's common sense knowledge into coherent and critical conceptions of the world. This is because everyday knowledge tends to be diffuse, uncoordinated, spontaneous and based on direct experiences. This form of learning demands a reciprocal engagement between theory and practice (Kontinen, 2013). It is concerned with internalisation and appropriation and can be achieved using artefacts or instruments that are already in place. The mature concepts can be later used as a basis for designing of new systems and instruments through knowledge generation and innovation. Sometimes an innovative solution emerges in practice but is not noticed as such before being applied in other context and discussed so that the new principle inherent in the practice is explicated. Sometimes research focuses on revealing the new principle that comprises the innovation.
- *Developing new knowledge and innovations:* This happens when a group of people with different experiences and perspectives work on the same object and jointly develop new knowledge or tools to address the new and emerging problems (Lave & Wenger, 1991). Collective learning involves the joint construction of what needs to be done, developing and implementing the creative solutions (Simovska, 2008). This third level of learning leads to the growth and development of the

> individual's knowledge and the transformation of practices, organisations and relationships. It involves both internalisation and externalisation, leading to expansive learning; and is based on an analysis of the origins of a phenomenon and experimentation that reveals functions and relationships between interacting things.

What has enabled human beings to continue to progress is the ability to accumulate culture and use it for the further development of human practices. The cultural transmission and creation processes are explained here by Tomasello (1999: p. 7) who notes:

> *Processes of cultural learning are especially powerful forms of social learning because they constitute both (a) especially faithful forms of cultural transmission (creating an especially powerful cultural ratchet) and (b) especially powerful forms of social-collaborative creativeness and inventiveness, that is, processes of socio-genesis in which multiple individuals create something together that no one individual could have created on its own ... Cumulative cultural cognition is thus the explanation for many of human beings' most impressive cognitive achievements.*

Cumulative cultural evolution on the other hand takes place through an iterative process of imitation, instruction and innovation. Through the creation of cultural tools and practices (socio-genesis) and cultural learning, human beings accumulate conceptual and material tools through historical time, causing a ratchet effect. Subsequent generations' cognitive development is built upon the entire cultural history of their social group (Tomasello, 1999). This is why cultural history is important for understanding learning, practices, human and societal development. If there is slippage in the accumulation process (such as has been the case in some oral traditions or marginalised cultures), cultural evolution is retarded (*ibid.*). This is especially relevant in the African context as O'Donoghue *et al.* (2014: p. 11) observe:

> *It is important to note that much of the African heritage landscape has been fragmented over many, many decades of colonial intrusion, exploitation, extraction, misrepresentation, marginalisation and modernising exclusion ... Much of the depth of indigenous wisdom in heritage practices have been uncovered, recovered, rediscovered through a slow process of local co-engagement and consultation with and among those indigenous people and representative who still have intergenerational artifacts, memory traces and stories to tell.*

1.6.2 Main outcomes of working with DWR

DWR is underpinned by three important principles, namely: ascending from the abstract – isolated, local and rudimentary like a seed, to the concrete – developed and enriched like a plant; double stimulation and transformative agency (Virkkunen & Newnham, 2013). The central method in DWR is called a formative workshop or CL workshop. In short, the DWR methodology enables:

- the development of new relationships of interaction, innovation and action involving people from different disciplines, performing different roles;
- the joint analysis, co-design and co-production of future-oriented solutions that address the causal mechanisms and help participants overcome current and emerging obstacles; and
- social transformation in concrete contexts of application through the integration of new solutions into practices and activities and the development of collective transformative agency.

1.7 Agricultural learning and innovation orientations that can benefit from DWR

Globally, there is a growing shift towards the development of agricultural learning, research, innovation and development practices that recognise and build on different ways of knowing. The four research, learning, innovation and development orientations that resonate with DWR thinking and intentions are:

- Integrated Agricultural Research for Development (IAR4D);
- development oriented agricultural research;
- people centred learning and innovation; and
- Agricultural Innovation Systems (AIS).

Some of the common features of the four orientations and DWR are:

- participation and collaborative partnerships;
- synergy, collective learning, innovation and action;
- productive activity; and
- demand-driven, purposeful change and transformation.

The four agricultural research, learning, innovation and development orientations are briefly discussed below.

1.7.1 Integrated agricultural research for development

FARA has begun promoting the IAR4D concept whose essence is to 'enable agricultural research to play more effective roles in catalysing development, by embracing a broader system of agricultural innovation that will facilitate interaction and enhance the flow of knowledge among all key actors in agricultural systems and value chains' (Adekunle *et al.*, 2013: p. 1). FARA is responsible for the African Union's New Partnership for Africa's Development's (NEPAD) agricultural research, technology dissemination and uptake. In its White Paper on IAR4D, FARA identifies the principles that underpin IAR4D (Hawkins *et al.*, 2009) as follows:

- Integrates the perspectives, knowledge and actions of different stakeholders around a common theme;
- Integrates the learning that stakeholders achieve through working together;
- Integrates analysis, action and change across different levels of environmental, economic and social dimensions of sustainable development to cover economic growth and development, natural resources conservation, and social inclusion and equity; and
- Provides for analysis, action and changes at different spatial (field, farm, watershed, etc.); economic (product, firm, value chain, business cluster, etc.) and social organisation levels (individual, group, community, organisation, innovation system, etc.).

It is also important to note that IAR4D is intended to create capacity changes at three inter-related levels: individual, organisational and institutional levels (*ibid.*)

1.7.2 Development-oriented research in agriculture

The International Centre for development-oriented Research in Agriculture (ICRA, 2007) defines the characteristics of development oriented agricultural research as involving:

- seeing knowledge generation as a means to achieve change, not as an end in itself;
- focusing on learning with others and learning how to learn;
- putting emphasis on innovation rather than research, people rather than technology;
- valuing and seeking a systemic understanding of the relationships between parts; and
- seeing participation as facilitating interactive learning and joint analysis, planning and action between and among collaborating stakeholders.

1.7.3 People centred learning and innovation

Some studies on the history and evolution of agricultural research and development have concluded that the approaches have changed over time based on how innovation comes about and the roles that are played by different actors in the process. The changes are also being accompanied by a shift from a focus on technology to innovation (Nederlof *et al.*, 2011). The current thrust is on People-Centred Learning and Innovation (Scoones *et al.*, 2008: p. 4). Some of its main features are:

- Scientists regard farmers as partners, collaborators, entrepreneurs, innovators; organised group setting the agenda, 'the boss';
- Farmers regard scientists as one of the many sources of knowledge and information;
- Farmers are co-generators of knowledge and information and negotiate with other partners in the innovation systems;
- The scope of innovation goes beyond the farm gate and covers multi-functional agriculture, livelihood, food systems and value chains across multiple scales from global to local and long time frames; and
- The innovators are multiple actors that come together to form learning and innovation alliances.

1.7.4 Agricultural innovation systems

Another important way of looking at the evolution of agricultural research and development is based on research systems. Table 1.1 gives a historical view of the evolution of agricultural research systems, which culminated in the emergence of AIS (Osiru, 2014). An agricultural innovation system is a network of organisations, enterprises and individuals whose focus is to develop new products, processes and forms of organisation into social and economic use (World Bank, 2006).

AIS has a mechanism called innovation platforms (IPs) that is used to support multistakeholder learning and innovation processes. IPs refer to a place of learning and change where multiple stakeholders with different backgrounds and interests work together to address a common goal, shared issue, challenge or opportunity so as to improve their livelihoods, enterprises and/or other interests (FARA, 2007). Nokoe *et al.* (2013: p. 473) go on to say 'An innovation platform refers to both the emergent properties of players and their processes, practices and habits, as well as the formal structures that might give operational focus to activities and interactions.' IPs are underpinned by systems thinking, which seeks to provide a holistic and comprehensive framework for understanding innovation as emerging from a broad

Table 1.1. History and evolutions of agricultural research systems (Anderson & Roseboom, 2013).

Defining criteria	National Agricultural Research Systems (NARS)	Agricultural Knowledge and Information Systems (AKIS)	Agricultural Innovation Systems (AIS)
Primary actors	research organisations	research, extension and education organisations	potentially all actors involved in the creation, diffusion, adaptation, and use of agricultural knowledge
Intended outcome	technology invention and transfer	technology innovation and adoption in agricultural production	technological and institutional innovations
Organising principle	using science to create new technologies	accessing agricultural knowledge	new uses of knowledge for social and economic change
Mechanism for innovation	technology transfer	knowledge and information exchange	collaborative interaction and innovation among stakeholders
Role of policy	resource allocation, priority setting	linking research, extension and education	enabling collaborative innovation
Nature of capacity strengthening	infrastructure and human resources development	communication between people	strengthening interactions between all actors involved in the creation, diffusion, adaptation and use of agricultural knowledge; and creating an enabling environment – ecologies for joint learning and innovation

network of dynamically linked actors (World Bank, 2006). At a local level such platforms tend to focus on improving practices through joint experimentation and linking farmers to markets. At national and international levels IPs tend to focus on broader contextual matters such as policy (Nederlof *et al.*, 2011). Figure 1.2 shows stakeholders who comprise IPs and the networked relationship between and among them as illustrated by the Sub-Saharan Africa Challenge Programme (SSA CP) (Anderson and Roseboom, 2013) in an IAR4D context.

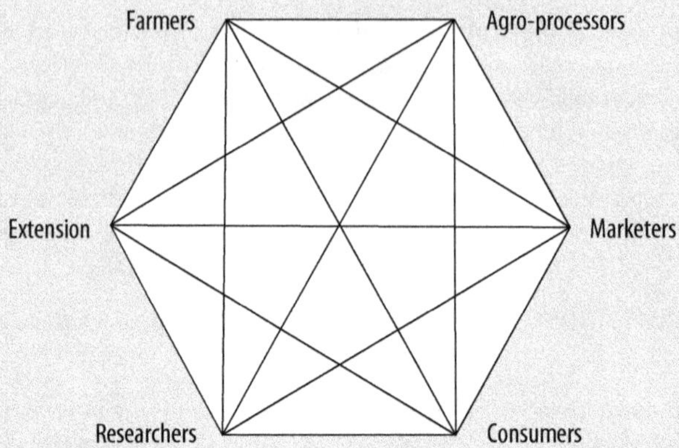

Figure 1.2. Interacting multistakeholder groups in an agricultural IP (Adekunle and Fatunbi, 2012, p. 987).

1.8 NRM approaches that can benefit from DWR

The discourse of sustainable development that gained currency from the 1980s following the Brundtland Commission of 1987 has shaped current approaches to NRM. The main contribution of the discourse has been to encourage the simultaneous tackling of the competing intentions of social justice, economic viability and ecological sustainability. The influence of sustainable development on educational approaches is partly reflected in the declaration of the United Nations Decade of Education for Sustainable Development (UNDESD). Its goal is to integrate values, principles and practices of sustainable development into all aspects of education and learning (UNESCO, 2005). Other streams of influence grew out of calls for civic participation in decision-making (Chambers, 1997) in NRM; and the rights-based discourse that underlined the importance of enabling the poor to access resources (Sen, 1999 cited in Cundill & Rodela, 2012).

1.8.1 Collaborative NRM

The above international discourses influenced the development of collaborative NRM to involve all key stakeholder groups in joint learning, innovation and action. Collaborative NRM produces the following outcomes:

- joint learning, which results in an expansion of the understanding of human-environment interactions and the problems at stake;
- joint problem solving;
- joint action; and
- conflict mitigation and relationship building (Cundill & Rodela, 2012; Roling, 2002).

The features of collaborative NRM that resonate with the DWR are:
- recognising the different values, interests and concerns of those living within and just outside protected areas;
- being open to various types of natural resources entitlements, and linking the entitlements to responsibilities in the NRM context;
- harnessing the complementarity of capacities and comparative advantages of different stakeholder groups; and
- learning by doing, through on-going revisions and improvements in NRM (Borrini-Fayerabend *et al.*, 2007).

1.9 The added value of DWR

Against the background of such a considerable range of approaches to stimulating agro-ecological innovations, it is necessary to identify the added value of DWR. This can be summarised as follows:
- It provides theory-informed methods for analysing developmental possibilities for interacting groups of people that have a shared concern.
- It is based on an understanding of the importance of conceptual and material tools in collective solution generation-based learning. When an insight is converted into an artefact, tool or sign that enables cultural learning, it stays in place for long enough to become a basis for future innovation (Tomasello, 1999: pp. 39-40).

1.10 Organisation of the resource book

The resource book is divided into the following four chapters:
- Chapter 1 primarily makes the case for developing the resource book;
- Chapter 2 provides the theoretical and conceptual framing of the resource book by discussing CHAT and the key concepts that underpin DWR;
- Chapter 3 illuminates the DWR process and the CL workshop method, which is central in the methodology. The chapter also discusses the associated

principles of ascending from the abstract to the concrete; double stimulation and transformative agency development. It draws on a good range of case studies to illustrate certain aspects of the DWR process; and
- Chapter 4 highlights some of the lessons learnt in utilising DWR in agriculture and NRM workplace contexts.

At the end of each chapter there is a list of references that is intended to direct the reader to relevant further reading.

1.11 Conclusions

This chapter has outlined the major reasons behind the writing of the resource book, drawing on the developments taking place in the agricultural sector on one hand and the collective, systems approach to learning and innovation on the other. It has identified some of the DWR-inspired interventionist studies in agriculture and NRM. It is worth noting that DWR is a methodology that is used in the broader field of social learning. It is a form of learning that contributes to, and occurs in a 'learning system' in which people with different backgrounds and vantage points learn from each other and collectively become more capable of withstanding setbacks and dealing with insecurity, complexity and risks as they jointly work towards improving their practices (Wals, 2012).

The list of research studies conducted using the DWR methodology curiously and largely focuses on different forms of ecological agriculture. This does not mean that the DWR methodology's use is limited to such practices. Instead, it reflects the kinds of dialectical issues of current human development aptly condensed in the concept of sustainable development. Creative tensions that serve as potential sources of new learning and innovation in agriculture and NRM appear to lie between: economic progress and ecological sustainability; long term interests and short-term interests on the other; and the distribution of costs and benefits along the value chain.

The history of agricultural research and development has shown a historical marginalisation of farmers and their contributions to the improvements of agricultural practices. This is the major reason why this book's primary entry point is and bulk of examples in subsequent chapters are based on the inclusion of farmers and farmer groups.

References

Adekunle, A.A., & Fatunbi A.O. (2012). Approaches for Setting-up Multi-Stakeholder Platforms for Agricultural Research and Development. *World Applied Sciences Journal 16 (7)*, 981-988.

Adekunle, A.A., Ayanwale, A.B., Fatunbi, A.O., Agumya, A., Kwesiga, F., & Jones, M.P. (2013). *Maximising impact from agricultural research: Potential of the IAR4D concept*. Accra, Ghana: Forum for Agricultural Research in Africa (FARA).

Adekunle, A.A., Ayanwale, A., Fatunbi, A.O., Olarinde, L.O., Oladunni, O., Nokoe, S., Binam, J.N., Kamara, A.Y., Maman, K.N., Danbegnon, C., & Emechebe, A. (2014). *Unlocking the potential for integrated agricultural research for development in the Savanna of West Africa*. Accra, Ghana: FARA.

Anderson, J., Roseboom, J., & Weidemann Associates (2013). *Towards USAID re-engaging in supporting national agricultural research systems in the developing world*. Washington, DC, USA: USAID Bureau of Food Security.

Belay, M. (2012). *Participatory mapping, learning and change in the context of bio-cultural diversity and resilience*. Unpublished doctoral thesis. Grahamstown, South Africa: Rhodes University.

Borrini-Fayerabend, G., Farvar, M.T., Nguinguiri, J.C., & Ndangang, V.A. (2007). *Co-management of natural resources: Organising, negotiating and learning by doing*. Heidelberg, Germany: Kasparek Verlag.

Chambers, R. (1997). *Whose reality counts? Putting the last first*. Rugby, UK: ITDG Publishing.

Chema, S., Gilbert, E., & Roseboom, J. (2002). *A critical review of recent experiences in reforming agricultural research in Africa: Draft Working Paper*. The Hague, the Netherlands: ISNAR.

Chima, J.M., & Chinkhuntha, G.J.C. (2002). *Transforming swamps and gullies into productive land: The case of Tikondwe Freedom Gardens of Malawi*. Harare, Zimbabwe: PELUM.

Cundill, G., & Rodela, R. (2012). A search for coherence: Social learning in natural resource management. In: H.B. Lotz-Sisitka, (Ed.),*Views on social learning literature: A monograph for social learning researchers in natural resources management and environmental education*. Grahamstown, South Africa: Environmental Learning Research Centre, Rhodes University/EEASA/SADC REEP.

Directorate General Development and Cooperation (DEVCO) (2014). *Research and innovation for sustainable agriculture, food and nutrition*. Brussels, Belgium: DEVCO.

Edwards, A. (2005). Let us get beyond community and practice: The many meanings of learning by participation. *The Curriculum Journal, 16 (1)*, 49-65.

Engeström, Y. (1987). *Learning by expanding: An activity theoretical approach to developmental research*. Helsinki, Finland: Orienta-Konsultit.

Engeström, Y. (1999). Innovative learning in work teams: Analyzing cycles of knowledge creation in practice. In: Y. Engeström, R. Miettinen & R. Punamaki, (Eds.), *Perspectives on activity theory: learning in doing: social, cognitive, and computational perspectives* (pp. 377-404). Cambridge, UK: Cambridge University Press.

FAO (2012). *Report of the FAO expert consultation on agricultural innovation systems and family farming*. Rome: FAO. Available at: www.fao.org/docrep/015/an761e/an761e00.pdf

FAO (2014). *The state of food and agriculture: Innovation in family farming.* Rome, Italy: FAO

Forum for Agricultural Research in Africa Sub-Saharan Challenge Programme (FARA/SSA-CP) (2007). *Security the future of Africa's children: Research plan and programme for impact assessment.* Accra, Ghana: FARA.

Gramsci, A. (1971). *Selections from the prison notebooks.* Edited and translated by Hoare, Q. & Smith, G.N. London, UK: Lawrence and Wishart.

Hall, A.J., Yogamand, B., Sulaiman, R.V. Raina, R.S., Prasad, C.S., Naik, G.C., & Clark, N.G., (Eds.) (2004). *Innovations in innovation: Reflections on partnerships, institutions and learning.* Andhra Pradesh/New Delhi, India: Crop Post-Harvest Programme/ICRISAT/National Centre for Agricultural Economics and Policy Research.

Hall, M. (2010). *Innovation in Africa: Speech on the Conference hosted by Trust Africa, CODESRIA and the United Nations Institute for Economic Development and Planning on 24 August 2010.* Available at: www.salford.ac.uk/_data/assets/pdf-file/0018/7X20/innovation-africa.pdf.

Hawkins, R., Heemskerk, W., Booth, R., Daane, J., Maatman, A., & Adekunle, A.A. (2009). *Integrated Agricultural Research for Development (IAR4D): A concept paper for the Forum for Agricultural Research in Africa (FARA) Sub-Saharan Africa Challenge Programme (SSA-CP).* Accra, Ghana: FARA.

Hill, R., Capper, P., Wilson, K., Whatman, R., & Wong, K. (2007). Workplace learning in the New Zealand apple industry network: a new co-design method for government 'practice making'. *Journal for Workplace Learning, 19 (6),* 359-376.

ICRA (2007). *Strengthening capacity for rural innovation: Building skills, changing outlooks, ICRA Annual Report.* Wageningen, the Netherlands: ICRA.

Kachilonda, D. (2014). *Investigating and expanding learning in co-management of fisheries resources to inform extension and training.* Unpublished doctoral thesis. Grahamstown, South Africa: Rhodes University.

Kontinen, T. (2013). A Gramscian perspective on developmental work research: Contradictions, power and the role of researchers reconsidered. *Outlines – Critical Practice Studies, 14 (2),* 106-129.

Lave J., & Wenger, E. (1991). *Situated learning: Legitimate peripheral participation.* Cambridge, UK: Cambridge University Press.

Leeuwis, C. (2004). Rethinking innovation and extension. In: H.A.J. Moll, C. Leeuwis, E. Mazungu & L.F. Vincent, (Eds.), *Agrarian institutions between policy and action: Experiences from Zimbabwe.* Harare, Zimbabwe: Weaver Press.

Leeuwis, C., Schut, M., Waters-Bayer, A., Mur, R., Atta-Krah, K., & Douthwaite, B. (2014). *Capacity to innovate from a system CGIAR research programme perspective. Programme Brief AAS-2014-29.* Penang, Malaysia: CGIAR Research Programme on Aquatic Agricultural Systems.

Lindley, D., & Lotz-Sisitka, H. (2014). *Expansive social learning, morphogenesis and reflexive action in an organisation responding to wetland degradation.* Paper presented at the 30th EGOS (European Group for Organisational Studies) Colloquium. July 2-5, 2014. Rotterdam, the Netherlands: Rotterdam School of Management, Erasmus University.

Machobane, J.J., & Berold, R. (2003). *Drive out hunger: The story of JJ Machobane of Lesotho.* Bellevue, South Africa: Jacana Books.

Masara, C. (2010). *Learning commercial beekeeping: Two cases of social learning in southern African community natural resources management contexts.* Unpublished Masters thesis. Grahamstown, South Africa: Rhodes University.

Mukute, M. (2010a). *Exploring and expanding farmer learning in sustainable agriculture workplaces.* Unpublished PhD thesis. Grahamstown, South Africa: Rhodes University.

Mukute, M. (2010b). *Improving farmer learning in and for sustainable agriculture in southern Africa. Gatekeeper 149.* London, UK: International Institute of Environment and Development (IIED).

Mukute, M. (2012). *Building farmer resilience by stimulating joint learning, collective and relational agency: the case of 32 organic farmer associations in Zimbabwe.* Paper presented at the Environmental Education Association of Southern Africa (EEASA) Conference. September 12, 2012. Grahamstown, South Africa: Rhodes University.

Mukute, M. (2013). Bridging and enriching top-down and participatory learning: the case of smallholder, organic conservation farmers in Zimbabwe. *Southern African Journal of Environmental Education, 29,* 74-92.

Mukute, M., & Lotz-Sisitka, H. (2012). Working with cultural historical activity theory and critical realism to investigate and expand farmer learning in southern Africa. *Mind, Culture and Activity, 19 (4),* 342-367.

Mukute, M., & Marange, F. (2011). *Mapping EU-SSA Agricultural Research for Development (ARD): Civil Society Organisations (CSOs) engagement and resource allocation processes.* Leusden, the Netherlands: ETC Netherlands.

Nederlof, S., Wongtschowski, M., & Van de Lee, F. (2011). *Putting heads together: Agricultural innovation platforms in practice. Bulletin 396.* Amsterdam, the Netherlands: KIT Publishers.

Nokoe, K.S., Van Rijn, F., Adekunle, A.A., Ayanwale, A.B., & Nyikahadzoi, K. (2013). Similarities among FARA-led IAR4D innovation platforms. *European Scientific Journal, 9 (36),* 472-484.

Nowotny, H., (Ed.) (2006). *Cultures of technology and the quest for innovation.* New York, NY, USA: Berghahn Books.

O'Donoghue, R., Shava, S., & Zazu, C., (Eds.) (2014). *African heritage knowledge in the context of social learning: Learning contributions of the Regional Centres of Expertise on education for sustainable development.* Yokohama, Japan: United Nations University, Institute of Advanced Studies.

Olvitt, L. (2010). Ethics-oriented learning in environmental education workplaces: An activity theory approach. *Southern African Journal of Environmental Education, 27,* 71-90.

Osiru, M. (2014). *Partnerships in agricultural innovation systems: An African researchers' perspective.* Bamako, Mali: ICRISAT.

Pereira-Querol, M.A. (2011). *Learning challenges in biogas production for sustainability. Studies in Environmental Science Volume 239.* Helsinki, Finland: Institute of Environmental Studies, Unigrafia.

Pereira-Querol, M.A., Seppaänen, L., & Virkkunen, J. (2014). Exploring the developmental possibilities of environmental activities: On-farm biogas production. *Environmental Science and Policy, 37,* 134-141.

Pesanayi, T. (2014). *Learning across socially-constructed boundaries: cases of expansive learning for resilience in agricultural and water sustainability practices in southern Africa*. Unpublished PhD proposal. Grahamstown, South Africa: Rhodes University.

Pimbert, M. (2009). *Towards food sovereignty: reclaiming autonomous food systems.* London, UK: International Institute of Environment and Development.

Pretty, J. (2003). *Social capital and connectedness: Issues and implications for agriculture, rural development and natural resources management in ACP countries. CTA Working Document Number 8032.* Wageningen, the Netherlands: CTA.

Radjou, N., Prabhu, J., & Ahuja, S. (2012). *Jugaad innovation: Think frugal, be flexible, generate breakthrough growth.* San Francisco, CA, USA: Jossey-Bass.

Roling, N. (2002). Beyond the aggregation of individual preferences: Moving from multiple to distributed cognition in resource dilemmas. In: C. Leeuwis, & R. Pryburn, (Eds.), *Wheelbarrows full of frogs: social learning in rural resource management* (pp. 25-48). Assen, the Netherlands: Koninklijke Van Gorcum.

Scoones, I., Thompson, J., & Chambers, R. (2008). *Farmer first revisited: Innovation for agricultural research and development, workshop summary.* Oxford, UK: ITDG.

Seppänen, L. (2004). *Learning challenges in organic vegetable farming: an activity theoretical study of on-farm practices.* Helsinki, Finland: University of Helsinki, Education Department.

Simovska, V. (2008). Learning in and as participation: A case study from health promoting schools (pp. 61-80). In A. Reid, B.B. Jensen, J., Nikel, & V. Simovska, (Eds.), *Participation and Learning: Perspectives on Education and the Environment, Health and Sustainability.* Bath, UK: Springer.

Stoop, W.A., & Hart, T. (2005). Research and development towards sustainable agriculture by resource-poor farmer in Sub-Saharan Africa: Some strategic and organisational considerations in linking farmer practical needs with policies and scientific theories. *International Journal of Agricultural Sustainability, 3 (3),* 206-216.

Tarulli D., & Cheyne, J.A. (2005). Dialogue, difference and voice in the zone of proximal development. In: H. Daniels, (Ed.), *An Introduction to Vygotsky (2nd ed.).* London, UK: Routledge.

Tomasello, M. (1999). *The cultural origins of human cognition.* Cambridge, MA, USA: Harvard University Press.

Triomphe, B., Waters-Bayer, A., Klerkx, L., Schut, M., Cullen, B., Kamau, G., & Le Borgne, E., (Eds.) (2014). *Innovation in smallholder farming in Africa: Recent advances and recommendations.* Proceedings of the International Workshop on Agricultural Innovation Systems in Africa (AISA). May 29-31, 2013. Nairobi, Kenya: CIRAD.

UNESCO (2005). *The United Nations decade of education for sustainable development: International implementing scheme.* Available at: http://www.unesco.org.

Vänninen, I. (2012). *Change Laboratory for supporting collaborative innovation and transformative agency in primary production.* Unpublished Masters thesis. Lappeenranta, Finland: Lappeenranta University of Technology, Faculty of Technology Management.

Virkkunen, J., & Newnham, D.S. (2013). *The change laboratory: A tool for collaborative development of work and education.* Rotterdam, the Netherlands: Sense Publishers.

Von Hippel, E. (2001). User toolkits for innovation. *Journal of Product Innovation Management.* Available at: www.evhippel.mit.edu/papers/section-5/.

Wals, A.E.J., & Bawden, R. (2000). *Integrating sustainability into agricultural education: Dealing with complexity, uncertainty and diverging worldviews.* Aberdeen, UK: University of Aberdeen.

Wals, A.E.J. (2012). *Shaping the education of tomorrow: 2012 full-length report on the UN decade of education for sustainable development.* Paris, France: UNESCO.

Wiggins, S., & Roepstorff, P.M. (2011). New global realities governing agribusiness. In: K.K. Yumkella, P.M. Kormawa, T.M. Roepstorff & A.M. Hawkins, (Eds.), *Agribusiness for Africa's prosperity* (pp. 24-37). Vienna, Austria: United Nations Industrial Development Organisation (UNIDO).

Wohlmuth, K. (2011). Strengthening technological effort and innovation capabilities. In: K.K. Yumkella, P.M. Kormawa, T.M. Roepstorff & A.M. Hawkins, (Eds.), *Agribusiness for Africa's prosperity* (pp. 165-199). Vienna, Austria: United Nations Industrial Development Organisation (UNIDO).

World Bank (2006). *Enhancing agricultural innovation: How to go beyond the strengthening of research systems. Economic Sector Work Report.* Washington, DC, USA: The World Bank.

World Bank (2008). *Agricultural innovation systems: From diagnostics towards operational practices.* Washington, DC, USA: The World Bank.

Chapter 2. The thinking behind Developmental Work Research

2.1 Introduction

The introductory chapter made a case for the use of DWR to strengthen collaborative learning, innovation and action in agriculture and NRM. This chapter discusses the main theoretical concepts that underpin DWR, which was developed in Cultural Historical Activity Theory (CHAT). This means that the methodology that is central to this book was developed within a particular theoretical framework (CHAT), whose building blocks are called theoretical concepts. CHAT in turn was developed based on particular assumptions about reality, which is what constitutes ontology. The ontology in which CHAT, and therefore DWR, was developed is called a *dialectical ontology of developing systems* (Virkkunen & Newnham, 2013), which integrates properties, relationships and movement. It has dialectical roots whose storyline is that systems develop through overcoming inner contradictions. This means that there is a hierarchical relationship between ontology, theory and methodology, with each shaping and underpinning the other respectively.

Having outlined the relationship between CHAT and DWR above, we move on to discuss the meaning of theoretical concept, which is central to this chapter. Virkkunen & Newnham (2013: p. 43) note that a theoretical concept:

> *... captures a system of functional relationships of complementarity and interaction that unites diverse objects and phenomena in a system and determines its change and development ... In order to determine the essential interactions, theoretical thinking has to consider the system historically in the process of its coming into being. A theoretical concept that discloses essential functional relationships makes it possible to assess the possibilities that have not yet been realised. Such concepts are expressed primarily in methods of intellectual activity like carrying out thought experiments, interpolating the development of current processes to the future and the like, but they are also expressed in models that serve as instruments in such an intellectual activity.*

A theoretical concept may also be seen as theoretically grasped practice (Engeström & Sannino, 2011). The primary interest of this chapter lies in explaining theoretical concepts in DWR to enable the reader to apply the methodology coherently in supporting innovation in agricultural practice. Having a deep understanding of the methodology also reduces chances of applying it mechanically, in ways that overlook or oversimplify social interactions and change.

2.2 CHAT and the ontology of developing systems

CHAT is underpinned by a special ontology of developing systems (Tolman, 1981). Textbox 2.1 explains ontology. The concept of *developing systems* resonates with the idea of innovation systems as discussed in the preceding chapter. The concept of systems recognises that a thing exists not just in itself but in terms of both its internal and external relationships, which are in a continuous state of flux (*ibid.*).

Understanding these internal and external relationships, which are essentially dialectical, forms the essence of understanding systems; and affords one the opportunity to support their development. In CHAT-informed DWR, activity systems form the basic unit of analysis. The systems are described as *developing* to refer to their motion and movement while at the same time implying a qualitative change or evolution from what was to what is – from the historical to the present system, state or practice. This qualitative change, which is based on the struggle of opposites – originates in contradictions, which arise from self-relation and relation to others – relationships and interconnections (Hegel, 1969).

Textbox 2.1. Ontology.

Ontology is concerned with philosophical questions about the nature of reality. It serves as both 'gatekeeper and bouncer' of a methodology (Archer, 1995) because how society is held to be shapes how it is studied. There are many assumptions about reality that have been put forward. For example, some philosophers say reality is three-layered, comprising the empirical, actual and real layers. Empirical reality is that which can be observed; actual reality is the second layer of reality and is what happens when events are activated. The real is associated with causal powers and structures (Sayer, 2000: p. 2). Others argue that reality is made up of substances and relationships (Emirbayer, 1997). The ontology on which CHAT and DWR are based assumes that reality consists of *dialectical processes* that are triggered by internal relationships, interactions and contradictions, and self-movement.

2.3 CHAT and dialectics

The philosophy that underpins CHAT, and the associated ontology of developing systems, is *dialectics* – an underlying philosophy of internal relations whose tensions and contradictions can result in-self movement of the system. Dialectics offers the possibility of explaining development through revealing inherent relationships and contradictions and identifying those that are fundamental to the movement of a system or process (Shirokov, 1937 cited in Tolman, 1981). Dialectics is best understood in the context of the *concrete* thing – which means understanding a thing in its interconnections and movements. For example, minorities have relationships with majorities and they do not exist without majorities and the very essence of minorities is determined by defining their relations with the majority (Tolman, 1981: p. 47).

Dialectical matters that have been identified in productive work systems and practices include tensions between the following areas:
- specific improvements and the whole system;
- doing the practice on the ground and reflecting on the practice 'from a distance';
- short-term, medium-term and long-term cycles of change; and
- what people are used to doing and producing and what they need to do and produce under changed circumstances (Virkkunen *et al.*, 1997: p. 158).

These insights point to areas where agricultural practice tensions are likely to be found, without limiting it to them.

2.4 Key concepts in CHAT and DWR

CHAT provides a framework for the study of learning and transformation processes in work activities, groups and organisations (Haapasaari *et al.*, 2014). It is based on the work of Lev Vygotsky (1896-1934) and his Russian colleagues Leont'ev and Luria who sought to overcome what they saw as the limitations of psychoanalysis and behaviourism. Vygotsky revealed that human learning, development and activity are *mediated* by technical and psychological tools that are available in the cultures. Human activity is therefore tool-mediated. The development, acquisition and development of new tools potentially increases people's sphere and level of influence (Engeström, 1987). Leont'ev's main contribution to CHAT lies in the theory of the hierarchical structure of human activity and the insight that individuals' actions are always taken in the context of a joint, object-oriented activity. The object of an activity system, which is always collective and not individual, can be given, projected or anticipated (Leont'ev, 1978). The object determines the kinds of goals and actions that can be pursued and serves as the motive force for driving the activity forward (*ibid.*).

Understanding the motives of individuals and groups involved in agricultural learning, development and innovation is therefore important in facilitating collective innovation and development. CHAT is informed by the classic philosophies of Kant and Hegel, Marx and Engels. For example, Kant posited that individual consciousness is formed under the influence of knowledge that has been accumulated by society and objectified in the world of things created by people. This underlines the importance of culture in knowledge acquisition or appropriation. Marx and Engels, drawing on the work of Hegel and Darwin, noted that people are not just a product of evolution and assimilators of culture. They are also creators and transformers of culture. This provides background to CHAT's idea of externalisation where people create new knowledge and transform structures, systems and practices.

Engeström and his colleagues (1987-2014) at the University of Helsinki in Finland have developed CHAT further. They developed the concept of interacting activity systems, which is critical in understanding and strengthening multistakeholder collaboration. CHAT's theoretical concepts that are discussed in the following sub-sections are:

- unit of analysis – individual action and cultural mediation; and collective activity systems;
- contradictions as sources of change and development; and
- cultural evolution and cultural learning.

2.4.1 DWR and unit of analysis

DWR was developed in the context of productive work with a view to continuously improving it. DWR works best in the context of productive work, which involves actors who carry out the work, conceptual and material tools that help the actors to carry out their actions; division of labour; and a community that utilises the goods or services. Such a productive system can be as basic as a farming family involved in horticultural production, dryland farming and small livestock keeping. It can also be complex fisheries involving fisher folks, aquatic ecologists, law enforcement agents, policy makers, fish buyers, fish processors and consumers. DWR has the potential to help productive systems, which are also called activity systems, to be collaboratively reviewed and continuously improved. The improvement of activity systems takes place when boundaries between actors are crossed; and when productive practices become better than they were before.

The ontology of developing systems has implications for the unit of analysis that is used in DWR. A unit of analysis defines the entity that is being studied, which can be something(s) or somebody/groups of people. Vygotsky (1934) defines a unit of analysis as a partitioning of the complex whole into units that possess all the basic characteristics of the whole – a vital and irreducible part of the whole. This is different from treating an element of the whole as the unit of analysis. The former is built on an appreciation of the thing and its connections and relationships; the latter is mechanical and based on the *thing-in-itself*. In DWR, the unit of analysis serves as a mediating conceptual tool for both the participating subjects and the researchers. It is used repeatedly to analyse and redesign various aspects of the participants' activity or practice (Engeström, 2011: p. 10). The conceptualisation of the most basic unit of analysis in CHAT, and therefore in DWR, has evolved over time from:

- Goal-oriented mediated action in first generation CHAT based on Vygotsky's work;
- Object-oriented, collective activity system in second generation CHAT based on Leont'ev's contributions. It grew out of the need to contextualise purposeful human action in society; to
- Shared object-oriented interacting (collective) activity systems in third generation CHAT, which was developed based on the work of Engeström and his colleagues. It grew out of the need for the theory to deal with cultural diversity and the need for dialogue between different perspectives and traditions (Cole, 1988). This is the most appropriate unit of analysis in multistakeholder agricultural innovation processes (Textbox 2.2). Each CHAT generation is explained below.

Textbox 2.2. Basic unit of analysis in an agricultural innovation system.

The basic unit of analysis in DWR is the collective activity system. An example of a collective activity is farming, which is collective in that it has many actors who play different but complementary roles towards achieving its societal role of providing food, raw materials, employment and ecological services. It forms the foundation of the interacting activity systems that are of interest to multistakeholder learning and innovation systems in agriculture and NRM. Although the first generation CHAT is no longer adequate as a unit of analysis, it is important to understand it as part of the background to the collective activity system. The anatomy of the three generations of activity systems is depicted through triangular models that were developed by Engeström.

This evolution of the basic unit of analysis is explained in the next paragraphs.

2.4.2 Human action and cultural mediation

The first generation CHAT is based on Vygotsky's work and focuses on the individual, whose learning, action and development is mediated by cultural tools. Its limitation was that it was individually focused. Vygotsky says that while both animals and people respond to stimulus, the difference is that people's responses are often mediated by technical and psychological tools that they have produced through their culture (Vygotsky, 1978). Figure 2.1 shows how Engeström represents Vygotsky's basic idea of how individual action is mediated by tools in the course of doing productive work.

An example of first generation activity theory in an agricultural setting could consist of:

- *Subject:* the farmer or resource user.
- *Tools:* indigenous and local knowledge, books, farm implements, seed and breed all of which carry in them the knowledge of previous (and current) generations who designed and selected them. These mediate the relationship between the farmer and the object.
- *Object:* land that is used for the production of food and raw materials.

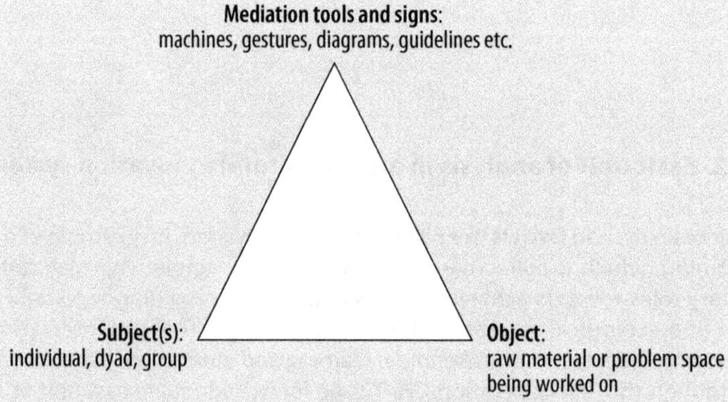

Figure 2.1. First generation CHAT.

It is important to note that a tool in one case can be an object in another as this depends on the structure of the action. For example, water can be as tool for an irrigation action but it can be an object in an action of water harvesting using such tools as a water harvesting tank, swales or dams.

Mediation entails going between, connection, linking or bridging. A mediator unites complementary and therefore contradictory elements into a functioning system (Textbox 2.3). For example, production and consumption are contradictory and complementary aspects of a whole that is tied together by the mediating function of a product. The ontology of developing systems entails the idea of analysis of dynamic relationships of interplay between complementary and contradictory elements. Sometimes mediation requires the use of both technical and psychological tools.

What is also important, and which led to the development of the second CHAT generation and related unit of analysis, is the realisation that human action takes place in a context. The context is characterised by community and societal motives to which the goals contribute; and by activities or practices within which actions are taken. The activity often means different things to different actors. There are different laws or rules that the actors take into account when they perform actions and there are different motivations or logics for individual human actors (Eskola, 1999). For example, the motivation of a plant breeder is different from that of a farmer and from that of an agro-dealer and from that of a consumer. This means that human action is informed by an interpretation of the situation and choice

Textbox 2.3. Mediation.

Mediation ... highlights human capacity to invent, build and use tools to transform our environment, as well as the capacity to create and use sign systems for meaning making and to develop as a result. Rather than acting on the world directly, our physical actions are mediated by the tools we build and use to work in and transform the environment. In addition, and more significantly, our experience in the world and our psychological representation of it is also mediated by the psychological tools we have created: speech and language, numerical and musical notations, cultural symbols and forms of art, and general semiotic systems among other things ... Using speech and language, humans initially learn to influence others, and ultimately learn to influence their own behaviour, including their own psychological action, as higher psychological functions develop (Vadeboncoeur & Collie, 2013: p. 208).

making. These interpretations and choices create tensions that need to be resolved. Table 2.1 shows three layers of causality of human action (Engeström, 2011: p. 40).

The first generation CHAT unit of analysis that is based on goal-directed mediated action is inadequate for interventionist research that is aimed at transforming communities and practices because it merely focuses on the individual, the tool and the product and neglects the context of the action, which provides the necessary relationships and connections. By extension, an intervention that is pitched at this level may be technically effective in the short run but is unlikely to last (Engeström, 2011).

For the developmental work researcher the main value of first generation CHAT lies in the mediation functions of tools and signs in productive work. Secondly, the concept of mediation helps one to overcome the dualism between the individual and its environment because mediation through tools and signs provides the linkage. Thirdly, the above triangle forms the most basic unit of analysis in CHAT but has been further developed under second and third generation CHAT, which are discussed next in this chapter.

2.4.3 Activity system

In CHAT, and therefore in DWR, *activity* refers to purposeful practice that is driven by a motive to meet a societal need. It is a socio-economic formation that is carried out by people with the aid of material and psychological tools (Engeström, 1987). Leont'ev commonly used hunting as an example of human activity. He identified the motive of hunting as to meet the food needs of society. Examples in Chapter 1 show that CHAT and DWR studies have focused on activities that include: vegetable farming in Finland; apple industry in New Zealand; sustainable

Table 2.1. Three layers of causality in human action.

Interpretive layer	In the actor	Takes into account according to this and that logic	If X, then Y rule, law
Contradictory layer	As participant in collective activities	Is driven by contradictory motives	Searching for resolution by often unpredictable actions
Agentive layer	As potential individual and collective agent	Takes intentional transformative action	Inventing and using artefacts to control the action from the outside

agriculture in Lesotho, South Africa and Zimbabwe; beekeeping in South Africa and Zimbabwe; fisheries in Malawi; bio-cultural conservation in Ethiopia; and wetland management in South Africa. In the case of agriculture, the societal motive can be viewed as to meet the food, fibre and nutrition needs of society and possibly to supply raw materials for industry. Within each activity, there are actions and operations (Leont'ev, 1978), with activity as the broadest concept. The table below (Table 2.2) summarises the hierarchy of human activity as understood in CHAT.

But Leont'ev (1978: p. 67) also makes an important point about the non-fixed nature of action and activity when he says:

> *Activity may lose the motive that elicited it, whereupon it is converted to action ... conversely, an action may turn into an independent stimulating force and may become a separate activity ... and finally an action can become a means of achieving a goal ... an operation.*

The distinction between an activity and an action is further made in the example below that was produced by Leont'ev (1981), one of the founding members of CHAT:

Table 2.2. Hierarchy of human activity (Leont'ev, 1978).

Level	Meaning
Activity	An activity is a concrete practice that is carried out by a community of people to meet a societal need. The object has two aspects: the given material reality such as agricultural land for the farmer; and the purpose to which the land will be put to serve a societal need. For example, the motive for an agricultural activity can be seen as to use land to achieve food and nutrition security in society. But in order to produce food, community members divide labour between and among themselves. In the field of agriculture, some people may produce seed, while others produce fertiliser, till the land and plant the seed, harvest, process, package and sell grain, legumes and vegetables. Their collective efforts comprise the activity.
Action	An action is conducted as part of an activity in the context of division of labour. The actions are determined by an individual's role in the joint activity. The actions are directed towards meeting a particular goal in a particular place and time. The goals are subjective and may not be directly related to the societal need or motive. For example, an agro-dealer's goal may be to make money while that of a researcher may be to publish papers and gain recognition. The goal may be seen as partial or intermediate results that contribute to the societal need. For example, producing fertiliser is an important goal which contributes to the production of food for society. Therefore agricultural activity comprises several interrelated chains of actions.
Operation	An operation is an action that has become an automated routine through rehearsal and repetition. Operations enable but do not determine the achievement of a goal. Driving a tractor, operating a rifle or typing letters using a computer keyboard are examples of operations that can be found in the agriculture and NRM activities.

> *When a member of a group performs his labour activity he does it to satisfy his needs. A beater, for example, taking part in a primeval collective hunt, was stimulated by a need for food, or perhaps a need for clothing, which the skin of the dead animal would meet for him ... It may have been directed, for example, at frightening a herd of animals and sending them towards other hunters, hiding in ambush ... The activity of this individual member ends with that. The rest is completed by other members. This result, i.e. the frightening of game, etc., understandably does not in itself, and may not, lead to satisfaction of the beater's need for food, or the skin of the animal. What the processes of his activity were directed to did not, consequently, coincide with what stimulated them, i.e. did not coincide with the motive of his activity ... We can say, for example, that the beater's activity is the hunt, and the frightening of game, his action.*
>
> *(Leont'ev, 1981: p. 210)*

Another interesting illustration of the difference between an activity and an action is given in the box below (Textbox 2.4). This is based on a more recent story that was told by an innovative farmer in Zimbabwe.

Box 2.4. An illustration of the difference between activity and action in CHAT.

In 2008, a farmer innovator in Chikukwa, Zimbabwe planted a lot of sweet potatoes and bananas at his new plot and failed to harvest anything because the crops were destroyed by moles. The moles also dug holes into the wall of his weir which he used for irrigating crops in winter, further undermining agricultural production. He started looking for a solution to the mole problem from 2009 and by 2013 he had developed an effective mole trap. In developing the tool, he drew on local inter-generational knowledge about hunting and made his own observations about the mole feeding habits (Mukute, 2014).

The activity that the farmer innovator was involved in is agriculture. The action that he took to improve his agricultural practice and be in a better position to realise the object of agriculture was mole trapping. The action of trapping and catching moles in itself does not produce crops. But when it is viewed in the broader context of the practice or activity, it contributes to agriculture in a similar way that the beater's actions contribute to the hunting activity described by Leont'ev above. The action of the farmer innovator reported here amounts to pest control in an agricultural activity. The operation in this action lies in the setting of the trap.

The story also serves to illustrate that an activity can become an action over time or in particular contexts. In this case the hunting of a mole, which could have been an activity associated with meeting societal need for food (the mole), has become an action in agriculture.

The concept of an activity described above forms the concept of an activity system, which is the basic unit of analysis in CHAT. It is based on the expansion of first generation CHAT by adding the community, rules and division of labour (Figure 2.2). An *activity system* is the minimal meaningful context for understanding individual action (Sawchuk, 2003). The main thing that distinguishes one activity system from another is the difference in the object (Daniels, 2001). The *object* refers to the concern, the problem space and carries the motivation and the meaning of an activity. The central relationship is between the subject and the object, which is mediated by conceptual and material tools. The subjects are often part of a community and their relationship with it is mediated by rules, while communities help achieve the system's outcomes through division of labour (Peal & Wilson, 2001). An activity system is heterogeneous and multi-voiced because subjects construct the object of the activity system in different and sometimes conflicting ways. The multiple voices arise from different perspectives which are informed by unique histories, interests and positions in the division of labour (Engeström, 2001). The elements of an activity system in CHAT, which mutually determine each other, are explained in Figure 2.2 and Table 2.3. Textbox 2.5 gives some illustrative examples of elements of an activity system in the context of agriculture.

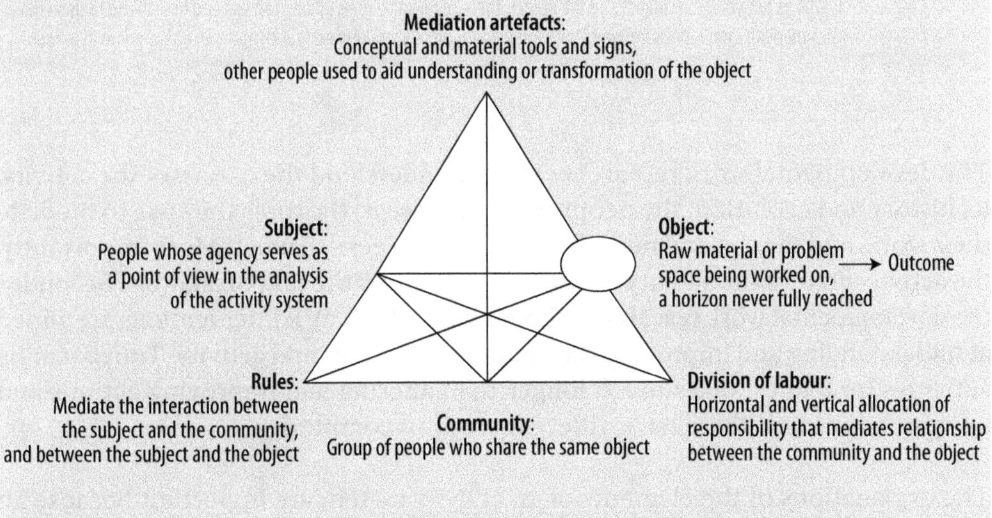

Figure 2.2. The structure of second generation human activity theory mode (Adapted from Engeström, 1987).

Table 2.3. Elements of an activity system.

Element	Explanation of the element
Subject	Individual or group of people whose agency is chosen as a point of view in the analysis of the activity system. The subject's relation with the object is mediated by four elements: rules, tools, community and division of labour, all of which carry cultural meaning.
Object	Raw material or problem space being worked on to produce something that meets human need (e.g. sick people in an area, or arable land). It is partly a given objective, material or ideal entity existing as given to the acting subject or anticipated, partly projected, planned, and constructed. It determines the range/horizon of possible actions. It also carries the purpose of the activity.
Outcome	Desired result of working on the object. It is also the specific value or values created through an activity. It carries the societal motive behind the activity.
Tools	Conceptual and material artefacts for understanding or transforming the object. Tools carry culture, history, skill and knowledge involved in developing them. An artifact becomes an instrument only when the subject uses it in dealing with the object (Virkkunen & Newnham, 2013: p. 34).
Community	Group of people who share the same object.
Division of labour	Horizontal and vertical allocation of responsibility which mediates relationship between the community and the object.
Rules	Rules mediate interactions between various elements of the activity system: between the subject and the community, as well as between the subject and the object. There can be also rules of who can use and how to use tools so that rules mediate interactions between individual and tools; community and tools; and tool and division of labour.

The developmental work researcher should understand the nature of the activity, its history and evolution, the people who practise it, the tools they use to mediate their work, and the societal need that the activity seeks to meet. More importantly, the activity defines the activity systems, which forms the unit of analysis. Secondly, the developmental work researcher should know that DWR interventions are aimed at understanding and improving the operations, actions and activity. Transforming activities (practices) takes much longer than altering and improving actions and operations because these have different levels of complexities.

The explanations of the elements of an activity system are important because an activity system forms basic unit of analysis in CHAT (Textbox 2.5).

Box 2.5. Illustrative examples of elements of an activity system.

People learn and do things through activities and for a purpose. People such as farmers, extension workers and researchers are the *subjects* in agricultural activities. Among other things, farmers, as subjects, are driven by the desire to produce crops and animals to eat, sell and feel secure about the future and to keep their land productive. This is the *object* of their work. In order for farmers to practise agriculture, they need resources such as seed, fertilisers and draught power. They also need the knowledge and skills to farm, which may be carried in their heads and hands, in their neighbours or in written forms. These different kinds of resources may be called *tools* even though some are intangible. Agricultural learning and practice is done in social, economic and ecological contexts. It is influenced by environmental factors such as soils, rains, snow and drought as well as by government policies, legislation and traditional norms. The natural and human-made conditions and provisions regulate how farmers work on their objects, are collectively called *rules*. Farmers relate with other people who also affect their business. Such groups of people may be buyers of produce, suppliers of inputs, scientists, policy makers, transporters and these make up the *community* in which farmers are the subjects. One of the important developments over the years that helped farmers to improve production is the sharing of roles in agriculture. For example, among family farmers, some members plant seed, others weed, and yet others harvest, transport or market produce. Sometimes people do the same duties together to make the work lighter or complete it sooner. This *division of labour* often helps farmers to farm better and can result in specialisation. All these parts interact together to form what is called an activity system, which can be represented in a triangle that show the connections and how the parts fit together (Figure 2.2). It is important to note that a tool, rule and form of division of labour are different kinds of cultural artefacts.

An activity system does not exist in isolation. For example, the family farm activity system or the farmer group activity system interacts with other activity systems. These include policy makers, agricultural extension workers, similar-minded civil society organisations, agricultural research organisations, agro-input suppliers and buyers. The development of a second generation activity system is more helpful when it reflects these external interactions. In this case, we call the family farm or farmer group activity system the central activity system (Figure 2.3). The activity systems that it interacts with are normally called stakeholders. A central activity system perspective is often helpful when the interventionist research is facilitating collaborative partnerships among people with different power relations. In this case farmer groups may need more support and accompaniment in preparing for interactions with partners. For example, Mukute & Lotz-Sisitka (2012) note that

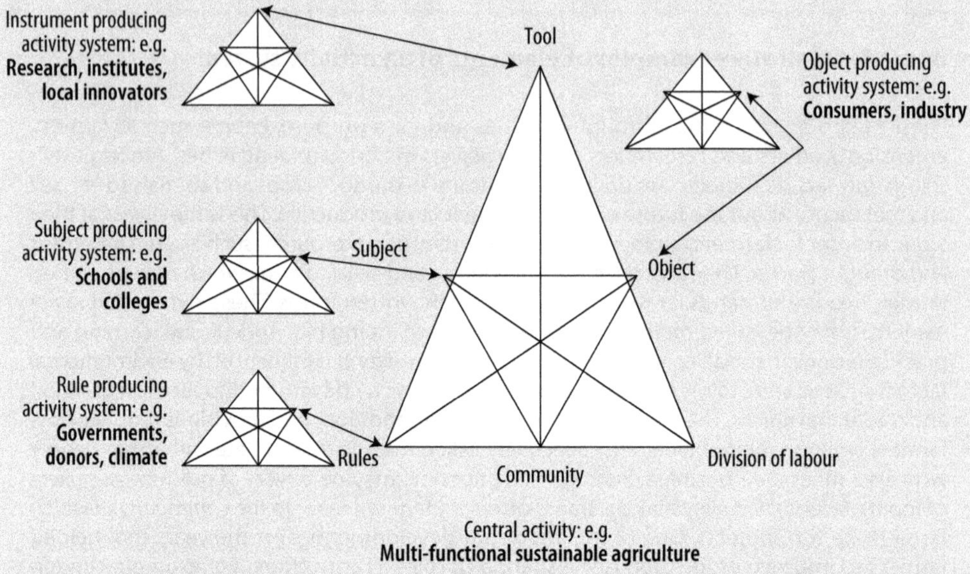

Figure 2.3. A central activity system of and its interaction with other activity systems (adapted from Engeström, 1987).

interventionist research involving smallholder farmers should pay attention to power relation issues as the farmers have to interact with:

- governments who have political power to decide on policies;
- the corporate sector with economic and cultural power to push for practices that increase their profits; and
- research organisations, universities and colleges that have intellectual or cultural power to choose what to teach.

The need for paying attention to power relations in agricultural innovation systems and processes has also been underscored elsewhere. For example, Lizares-Bodegon *et al.* (2002: p. 27) conclude that:

> *Caught in the clutches of poverty, inequality and political opportunism, farmer organisations are usually handicapped in [Participatory Technology Development] PTD negotiations and arrangements. In these platforms they face more articulate, more powerful and better resourced entities such as*

> *research institutions, NGOs and other civil society actors ... The feasibility of PTD efforts depends largely on how farmers are catalysed and empowered by the vision and the workings of their organisations.*

In Figure 2.3, the smaller triangles represent other activity systems or stakeholders that have a bearing on the central activity system. For example:

- governments may represent the rule-producing activity system;
- research institutes or local innovators may represent the tool activity system;
- the education system may represent the subject producing activity system; and
- consumers of agricultural produce may represent the object-producing activity system.

In working with this and the next unit of analysis, the developmental work researcher is encouraged to remember that:

- The activity system has a history and evolves over time and the diagrams merely give a 'frozen' picture of the rather dynamic activity system. The motion and associated development of an activity system, which a central idea of 'developing systems' is important to bear in mind in the research process even though it is not captured in the diagrams and associated tables.
- The best way to establish the special character of an activity system and its elements is to analyse the historical development of the activity and to compare older and newer forms of its elements.
- The elements of an activity system reciprocally interact and influence each other.
- The activity system is an *open* and developing system that interacts with neighbouring activity systems in a manner that results in interpenetration. For example, the tools that are adopted from other activity systems have to be adapted for use in the central activity to suit its social ecological conditions. Externally given rules are always interpreted in a specific way in the activity before they start to govern actors behaviour and practices.
- The meaning of the model is not primarily to describe an activity as it is now, but to analyse the activity in its development and to identify current inner contradictions in its structure due to changes in and between its elements.

2.4.4 Interacting activity systems

A more complex unit of analysis, which resonates with interactive collaboration among multistakeholders, is two or more interacting activity systems, which define third generation CHAT. Such interacting activity systems should have a (partially) shared object, which drives their collaboration. It resonates with current thinking on innovation systems in agriculture and NRM as discussed in Chapter 1. The most basic form of a shared object between two interacting activity systems is represented in the figure below (Figure 2.4). *Boundary crossing* is a central idea, which encourages actors from the different activity systems, working towards a shared object, to move into unfamiliar territory and jointly develop new solutions. For example, Hill *et al.* (2007: p. 364) identifies the shared object of the New Zealand government and the New Zealand apple industry as 'A profitable, sustainable legal system of production'.

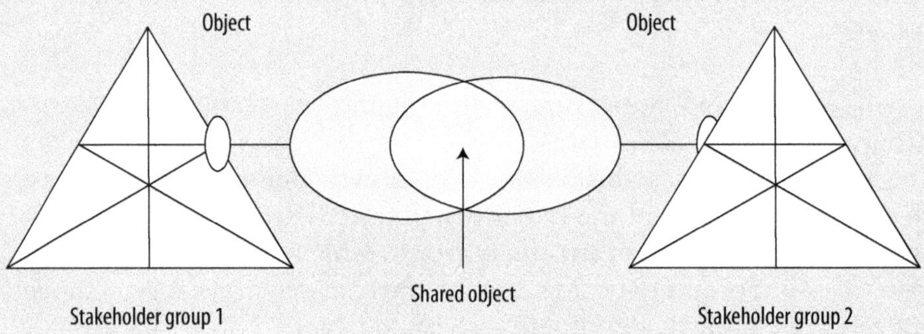

Figure 2.4. Most basic structure of third generation CHAT (Engeström, 2001).

2.5 The development dimension of activity systems

The above sub-section discussed the unit of analysis and how this has evolved in CHAT and DWR. This sub-section seeks to illustrate the 'development' or motion dimension of activity systems to reveal their evolutionary nature based on concrete examples. Qualitative changes in practices or activity systems are the essence of development. When viewed in the context of productive work, development can be understood as the general historical process of socialisation of production: enriching productive work with more intensive use of knowledge science and technology, and expanding the spheres of exchange and communication. Socialisation of production in this case takes place when productive work embodies the capabilities of the larger society rather than only those that emerge from isolated, local context (Adler, 2012). For example, the infusion of refrigeration knowledge and technology into the Mitticool illustrates the externalisation of both local and wider society knowledge.

2.5.1 Zone of proximal development

Zone of Proximal Development (ZPD) is the metaphor that is used in CHAT to refer to the next and higher form, level or stage. It is reached through experimentation, re-mediation and theoretical generalisation (Engeström, 2008). Vygotsky (1978: p. 86) defines ZPD as: '... the distance between the actual developmental level as determined by individual problem solving and the level of potential development as determined by problem solving under adult guidance or in collaboration with more capable peers'. The ZPD 'is a zone of human development, the frontier where we can find the links between the situated-embodied mind and the cognitive mind; the individual mind and the social mind; the development already attained and the development to be attained' (Del Rio & Alvarez, 2007: p. 301). But then, the primary focus of development in CHAT is not the individual person but the 'collective'; and associated learning environments or ecologies comprising multiple actors such as farmers, extension workers and researchers; agricultural land, agricultural colleges and research institutes. In this context, Engeström's concept of the ZPD is helpful – it occurs when the recurrent double bind situations in individuals' daily actions in an activity can be overcome by collaboratively creating a historically new form of the activity that has become culturally possible (Virkkunen & Schaupp, 2008). Simovska (2005, 2008) points out that this view of ZPD provides deeper consideration of the interaction between *cognition, context and practice* thus making learning and development more relational and socio-cultural (Simovska, 2008: p. 64).

2.5.2 Remediation

Remediation is concerned with transforming a practice or activity through a specific sequence of actions of knowledge creation by actors involved in the activity system (Virkkunen & Newnham, 2013: p. 48). It comprises three layers: the interpretive layer, the contradictory layer, and the agentive layer (Engeström, 2011) which put together comprise the key stages of a DWR journey. Remediation often leads to the formation of new forms of activity, with different or more complex objects. This therefore means that remediation is an important part of the development of activities systems from old to new. Engeström (2008: p. 11) explains remediation as follows:

> *The creation of new activity is a process of reflective re-mediation. A mediating concept or device can come up with a new object and a new activity. This kind of re-mediation is radically different from goal-rational theories of change. The limitation of the goal-rational models of creation and change is that they require*

> *that the investigator or the interventionist defines the desired outcome of the change effort at the beginning. This leads to a paradox: How can you create something new if you know ahead of time what it is? Re-mediation involves a shift from predefined or 'given new' goal to an unexpected or 'created new' object.*

Put differently, remediation can be of two kinds: one that results in a better understanding of a situation and is based on the relationship between the subject and the object. For example, a new pair of spectacles would enable the user to see better and see what he/she could not see before. The second and related meaning is concerned with the remediation of a contradiction.

A practical example in southern African agriculture is how to deal with the increased frequency and severity of droughts in the context of climate change. The 'created new' object is human-induced climate change.

There is therefore a close connection between ZPD and remediation: when the collectively defined ZPD is achieved, and a new one formulated, remediation would have taken place. Below is an example (Figure 2.5) of how the two concepts were captured in a southern African agricultural research study that was anchored on developing systems (Mukute, 2010a: pp. 370-377):

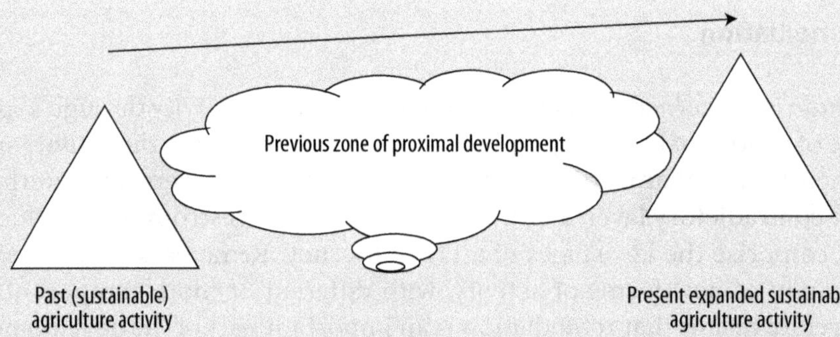

Figure 2.5. The zone of proximal development of sustainable agriculture (Mukute, 2010a).

> *Between Kwame Nkrumah's vision of an independent Africa and the realisation of democracy in South Africa in 1994 as the last African country to become independent, lies the zone of proximal development for an African renaissance ... a zone defined by a collective, not an individual.*
>
> *It is year 2072, exactly 100 years after the famous and first World Conference on Environment and Development; 93 since the Peasant's Charter at the World Conference on Agrarian Reform and Rural Development; 80 after the Rio Conference and 70 after the World Summit on Sustainable Development ... Between the surge of sustainable agriculture and the fall of conventional agriculture, lies the zone of proximal development of the dominant form of agriculture today. Within this space, several, multi-layered transformations happened in different places and at different times, all of which were coordinated towards an agriculture that produces enough safe and nutritious food; ... economic viability for the farming populations and professionals; ... improved farms, ... whose productive potential was multiplied and ecological services enhanced alongside improved access to land by the formerly poor and marginalised, better distribution of costs and benefits of agriculture ... It has made vast progress in capturing and sinking the notorious excessive human-generated carbon, which induced climate change ... This agriculture is taking care not to pollute the biological diversity of neighbouring ecological systems, not to pollute water bodies ...*
>
> *In the past, ... the real attention went to the now discredited 'conventional agriculture,' which produced high yields ... made the rich, richer; impoverished the soils; contributed to global warming and polluted water bodies and increasingly marginalised the farmers and other producers, as the corporate sector took over more and more control of the agricultural production chain ... Isn't it amazing – how the thoughts and deeds of the long dead live in us, in our practices and tools, as will our thoughts and deeds of today live on in the practices and tools of the future?*

For the developmental work researcher, remediation is a critical concept because it sums up the steps that ought to be taken collaboratively in the process of deliberately developing practices. But the scope of the ZPD can be much wider than this, especially when it involves the creation of a new theoretical concept such as the development of a steam engine was when it introduced. The new principle can, and in the case of the steam engine, was applied to a wide range of activities.

2.5.3 Examples of developing agricultural systems

There are numerous examples that can be given to illustrate how agriculture and NRM have evolved over time and how this has been manifested in the nature of mediating tools or devices; or that of objects as discussed by Engeström (2008) above in relation to motion, development and remediation of practices or activities.

2.5.3.1 Generations of evolving agricultural mediating tools

Figure 2.6 shows the goal-oriented action of producing crops and how this has changed over several cultural layers (generations) in a southern African context. The different forms of fertilisers are a form of material or technical tool that empowers the human being to act on the world more effectively. The human action of applying fertilisers constitute part of a set of actions that result in an agricultural activity whose object could be the concrete farm that is used for the purpose of meeting society's food and nutrition security. Some of these actions are ploughing, sowing seed, weeding, harvesting and processing. It is based on a crop production system in which the nutrients available in the farm are inadequate to meet the nutritional requirements of crops grown.

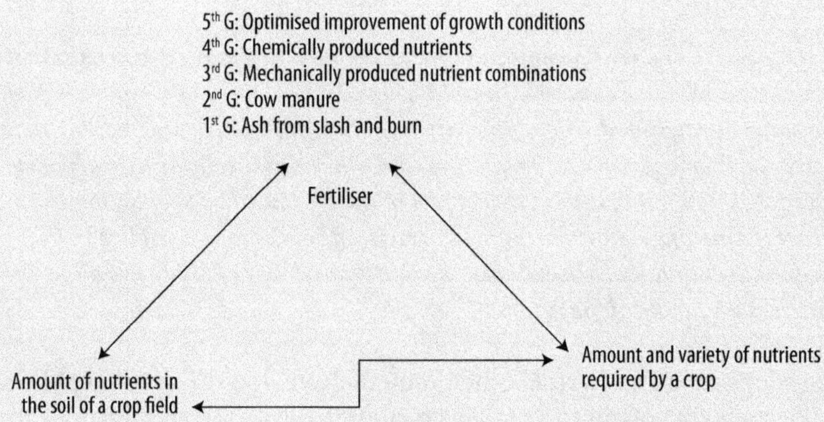

Figure 2.6. The tools and actions that farmers developed and took to improve crop production over time (G = generation).

Crop production can also be mediated by agricultural knowledge that can be represented by psychological tools such as guidelines, diagrams, formulae and designs. These are essentially representations and signs that people have culturally developed in order to help them understand, cooperate, communication and shape how they work with the technical tools. In other words, psychological tools are used for mastering and controlling people's own thinking, behaviour and social interactions. Such 'reflexive embodiments' according to Marx Wartofsky, can be gestures, visuals or oral.

2.5.3.2 Generations of evolving agricultural objects

The objects, concerns or problem spaces of agriculture as productive work have changed over time, becoming more and more complex. During the pre-industrial period, people farmed to meet their food needs. But when the industrial period emerged, accompanied by urbanisation the farmers produced crops to feed people, agro-based industries, and even to feed domestic animals. This generated income through sale of produce and through employment on farm and in industry. But the intensive utilisation of land, application of chemical fertilisers and poor land husbandry created other challenges that gave rise to the need for farming responsibly. This led to the emergence sustainable agriculture practices, which have the added object of protecting and enhancing the agro-ecological base and producing ecological services. Figure 2.7 captures the development of different and more complex objects, which have continuously redefined activity systems and their elements. This has culminated in the necessity for working with interacting activity systems and multistakeholders.

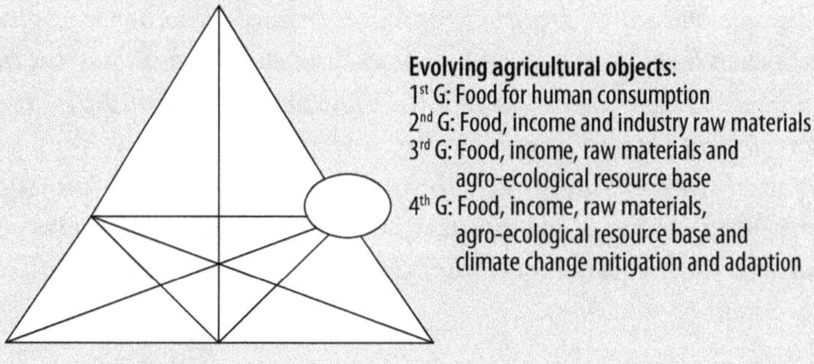

Figure 2.7. Changes of the objects of different forms of agricultural activities.

2.6 Cultural evolution and learning and their effect on practice

Cultural evolution and learning are central concepts in DWR as they inform how the methodology ought to be worked with. Cultural evolution acknowledges that when an intervention is made to improve a practice, there is already a cultural heritage on which improvements and changes are to be built. Cultural learning on the other hand provides ideas on the ways in which learning is viewed. What is particularly important in this case is that cultural learning transcends top-down and bottom-up learning.

2.6.1 Cultural evolution

Cultural evolution refers to the way in which human cultural traditions and artefacts are transmitted and accumulate from one generation to another. Vygotsky and Luria note that people are different from other animal species because they develop through culture and pass it on to succeeding generations, thus linking the past (history) and the future (Cole & Gagdamaschko, 2007). These inter-generational linkages help cultures to historically evolve. However, disruptions can lead to regression. This means that cultures, practices and their tools have histories. Tomasello (1999: p. 5) explains cultural evolution as follows:

> *Basically none of the most complex human artefacts or social practices – including industries, symbolic communication, and social institutions – was invented once and for all at a single moment by one individual or group of individuals. Rather, what happened was that some individual or group of individuals first invented a primitive version of the artefact or practice, and then some later user or users made a modification, an 'improvement,' that others then adopted and perhaps without change for many generations, at which point some other individual or group of individuals made another modification, which was then learned and used by others, and so on over historical time in what has been termed 'the ratchet effect' ... The process of cumulative cultural evolution requires not only creative invention but also, and just as importantly, faithful social transmission that can work as a ratchet to prevent slippage backward – so that the newly invented artefact or practice preserves its new and improved form at least somewhat faithfully until a further modification of improvement comes along.*

Cultural evolution therefore captures the entire collective wisdom of an entire social group throughout its whole cultural history (*ibid.*). It depends on two processes: innovation on one hand; and appropriation/internalisation on the other – one following the other over time. The cultural historical nature of human activity also means that activities are culturally layered: there are old and new elements in them, which makes history not only important as a narrative but also for the analysis of the historical origin of the elements of current activity systems – and of the activity system as a whole. In CHAT, culture refers to the entire pool of artefacts, including language, speech signs, norms, customs, tools and values accumulated by the social group in the course of its historical experience (Cole & Gagdamaschko, 2007). Some cultural products are used for the development of new and more sophisticated tools to further mediate human beings' ability to master their behaviour and nature. Human beings' development occurred along two dimensions: the increased use of artefacts in place of acting directly on nature; and the increased complexity of the meditational means themselves (Cole & Hatano, 2007: p. 111).

The developmental work researcher should make time to understand the historical development of the practice and the community that he/she is working with. This can be achieved through research methods such as ethnography, document analysis, and historical analysis. This helps in the identification of historically emerging challenges, obstacles and limitations as well as their underlying causes. On the other hand, it helps him/her to identify innovations and tools that have potential for use in the future in the context of an appreciative inquiry. Getting a historical perspective complements a systems perspective that a developmental work researcher gets from an analysis of the current and existing realities and relationships. Textbox 2.6 looks at cultural evolution in the context of intergenerational learning in African agriculture while at the same time revealing some of the problematic issues.

2.6.2 Cultural learning

Human *cultural learning* involves imitative, instructed and collaborative learning (Tomasello *et al.*, 1993). It is enabled by people's ability to see others as intentional and mental beings like themselves and put themselves in the 'mental shoes' of the other. This enables them to *learn from* as well as *through* the other because cultural artefacts and social practices such as tools and linguistic symbols respectively point beyond themselves. For example, tools point to problems they are designed to solve and language symbols point to communicative situations they are designed to represent. Therefore, learning about the tool or symbols involves learning about the purpose to which they are to be put – that is, the intentional significance of the tool (Tomasello, 1999: p. 6).

Imitative learning usually involves children learning from and through others by imitating adults' intentional, goal directed action. This may be concerned with

Textbox 2.6. Intergenerational cultural learning in African agriculture.

Education and learning, both formal and informal, is the basis of acquisition, sharing and transfer of knowledge and skills to enhance equity among the different generations ... In African agrarian communities, local and indigenous knowledge systems have been central in traditional agriculture and farming practices; and intergenerational learning was strongly embedded in cultural values, agricultural practices and farming activities such as planting, weeding, crop harvesting and ... animal husbandry. In the colonial and post-independence era, formal education has gradually replaced the traditional modes of education and learning among agricultural communities, but has not sustained the traditional intergenerational nature of knowledge acquisition and sharing. Modern formal education has thus done very little to generate a positive attitude towards agriculture among young people in Africa even though agriculture has and will continue to play a substantive role in their lives ... An intergenerational approach can add value to the current education, learning and knowledge acquisition and sharing processes in African agriculture ... The role of technology cannot be gainsaid. For example, ICTs can be critical in creating a joint and mutual intergenerational mentorship process whereby the technology savvy youth can mentor the older farmers in the use of new technologies ... In practice, this approach will require the explicit integration of intergenerational perspectives in policies, programmes, initiatives and processes that seek to increase the contribution of young people to agriculture in Africa pursued by different stakeholders along the agricultural value chain.

Source: Osano & Adam (2014, pp. 164-165).

how to use artefacts and gestures. Instructed learning is most common in school and college settings. In both forms of learning, it is important that the learner acquires the right knowledge and skills in order to avoid slippage, going backwards. Collaborative learning takes place in two ways: through the *ratchet effect* that takes place through virtual collaboration between generations; and through *joint collaboration* between two or more individuals working on the same problematic situation. Virtual collaborative learning takes places along a historical timeline when an inherited tool or practice is modified or improved to meet prevailing conditions, opportunities or challenges. Simultaneous collaborative learning takes place through processes such as Innovation Platforms discussed in Chapter 1, where dialogic interaction takes place (Tomasello, 1999). Therefore, we can conclude that learning practices entails:

- increased individual and collective understanding of the situation based on the appropriation of existing knowledge;
- the collective creation of new knowledge and practical solutions to concrete issues that leads to an improved (vertical movement) practice; and

- the integration of the new knowledge and solutions in everyday practices (horizontal movement).

Bateson's levels of learning (Textbox 2.7) provide important insights into both cultural learning and cultural evolution.

Textbox 2.7. Bateson's learning levels and their link to human action.

Bateson (1972) developed three layers of learning levels that are closely aligned to the hierarchy of human actions: operations, actions and activity as interpreted by Engeström (1987).

Learning level I is associated with learning *operations* through habituation, conditioning and rote learning in circumstances where the learning outcome and instruments of learning are given (Bateson, 1972). Examples include learning how to drive, read, write, till the land, dry vegetables, make a chemical solution using an existing formula or make nets for fishing.

Learning level II is concerned with *performing actions* to meet goals by solving problems in a situation where the outcome is given but the instrument has to be found through experimentation. It results in the formation of theoretical abstractions and novel solutions (innovations). For example, the farmer innovator who developed the mole trap intended to address a mole pest problem, which had no effective existing solution. The existing object that he was trying to address was household and community food security. As part of experimentation, he tried vertiver grass, fertility trenches and available rodent traps. He finally produced an effective trap after about four years of experimentation and the solution was adopted by the farming community in his area.

Learning level III is directed at a societal network of activities and takes place where both the instruments and the outcomes of learning have to be found or invented. It is motivated by the resolution of societal contradictions that go beyond the individual and are addressed through collective activity, which results in the transformation of contexts, philosophy, ethics, individual world outlooks and practices. This helps the learner to direct his or her learning level II and is often accompanied by the development and application of new methodologies. For example, the current sustainability world view, illustrates learning level III. It emerged in response to unintended negative and double-bind situations created by modern development and took several decades to form. It is shaping how people think about their environment and relate with it and with one another. The intended outcomes of sustainability are prosperity, social justice and ecological sustainability. Practices that are associated with this world view include sustainable agriculture and collaborative management in NRM.

Similarly, in a discussion on hierarchy of tools, Wartofsky (1979) identifies three levels of tools used in human activity: primary tools that are used directly in the productive activity (such as farm implements and seed in agriculture). Secondary tools are representations of practices and include agricultural production methods, guidelines and books. Tertiary tools are used for evaluating and developing secondary tools.

2.6.3 Learning and practice in the context of CHAT and DWR

Learning entails making the right choices from a given range of choices to perform tasks in the course of productive activity (learning the current practice); and expanding the variety of possible choices through innovations (incremental improvement, and transformation of practice). These three types of learning associated with CHAT and DWR are explained below:

2.6.3.1 Learning the current practice

Learning the current practice is associated with mastery of available knowledge and skills, which results in the effective performance of tasks in the practice. Such learning is often aided by other people such as subject specialists and fellow practitioners, who help the learner to appropriate the necessary knowledge or to identify the critical factors that affect the execution of the task. In this form of learning, both the object or outcome of learning and the instrument of learning are given. This level of learning is associated with internalisation. For example, an experienced and knowledgeable farmer who wants to train fellow farmers about how to select seed already knows that intended outcome of the learning as mastering seed selection and also the means by which he/she will train fellow farmers to acquire the necessary skills. Another key aspect of this learning level is that it tends to be focused on the individual. There is also often a clear and fixed way of carrying out the task or operation by an individual. Learning new things in this way can take place within hours. Learning the current practice enables the learner to appropriate knowledge from culture, understand the world better and to act upon it in potentially more diverse ways. Such knowledge also forms an important basis for developing further learning by the learner.

2.6.3.2 Learning related to incremental improvement of current practice

This is associated with improving actions towards goals, which are a sub-structure of a practice or activity. The learner makes sense of empirical data or existing knowledge by *reproducing* it in a processed form using instruments that are already existing in culture. This could lead to the formation of 'mature concepts'. Then, based on this higher level of understanding, the learner works with others to produce a concept, tool or instrument that improves a practice without changing its object. In a sense, the learning process is *productive* and involves externalisation. It is likely to be pursued by actors in one part of the activity system. Sector-based improvements in practices that are cross-sector constitute an example of this level of learning. Learning of this kind takes place in months and years.

2.6.3.3 Learning related to the transformation of practice

Learning concerned with the transformation of practice is learning that generates knowledge and innovations that result in substantial changes in a practice, including the expansion of its object. This results in what Engeström (1987) calls a *historical increase in the numbers and interconnections of human activities and the widening of the object-field of those activities.* It takes place when the contradictions between the new activity system and of those with which it interacts are resolved through their restructuring. It therefore tends to involve multiple interacting activity systems. In this way, the context of the problem is broken, altered or expanded. The movement, which is both vertical and horizontal, may also result in the development of new world views or paradigms (Textbox 2.8). The process of transforming practice takes several years.

Another interesting way of looking at both cultural evolution and learning is to see how human teaching and learning has evolved over generations, which suggests the following changes in emphasis (Donald, 2001):
- *Episodic* – accidental teaching based on short-term immediate needs;
- *Mimetic* – showing and correcting based on learning how to make tools and products from others;
- *Mythical* – story telling, metaphors and interpretation based on paintings and sculptures; and oral and written stories of human's interpretations of the world;
- *Theoretical* – anticipating and preparing based on theoretical concepts abstracted from human experience in the world; and
- *Participatory* – future and shared object-oriented collaboration and co-creation based on collective consciousness and intelligence.

Textbox 2.8. Changing paradigms about the ecological environment.

The world's view on the ecological environment has been changing over time as people's knowledge and wisdom about it has grown and changed. Martinez-Alier (2002) describes three paradigms that have evolved over time: the cult of wilderness, eco-efficiency, and environmentalism of the poor respectively. These paradigms have also had an effect on how natural resources utilisation is viewed. The cult of the wilderness is concerned with the preservation of pristine areas against growing human population and industrialisation. It seeks to insulate such areas from physical exploitation by human being through strategies such as national parks, world heritage sites, and identifying endangered species. Eco-efficiency seeks to manage and reduce the negative impacts of urbanization, industrial activities and agriculture on the environment. It recognises the *interaction* between the protected areas and the other areas and is based on the thinking that technological development can be used to reduce the negative effects on the environment alongside the use of taxes to reduce environmental pollution.

Environmentalism of the poor is primarily concerned with social justice in benefiting natural resources and distribution of negative effects of modern development. It seeks to ensure that the poor communities and countries benefit from their natural resources and that the rich communities and countries whose activities produce ecological 'bads' such as climate change bear the costs arising from their actions (*ibid.*). This shift from viewing technological solutions as the source of solutions to looking for solutions beyond them can be linked to the shift from seeing experts as the only source of knowledge to seeing laypeople as important sources of knowledge and solutions as well. This is what Visvanathan (2006) calls *cognitive justice* and it forms an important lens for viewing cultural evolution and learning as viewed in this book. These paradigm shifts have also shaped what is taught, learnt and practised in NRM and agriculture while at the same time defining the spaces occupied by different actors in knowledge and innovation development and use. For example, in agriculture environmentalism of the poor is associated with a shift from food security to food sovereignty, and from 'industrial' agriculture to sustainable agriculture (McMichael, 2014).

2.7 Contradictions and practice

It is important to emphasize that DWR does not merely seek to understand the situation but to improve it by identifying and addressing historically accumulating obstacles as well those that emerge in the process of implementing innovative solutions generated through the interventionist research. The structural factors that give rise to the obstacles, challenges and tensions that people and practices experience are called *contradictions* (Engeström, 2001). Ilyenkov (2007) notes that human and societal development is achieved through the identification, clarification and resolution of contradictions. And so do practices.

2.7.1 Using contradictions to develop and expand practice

Ilyenkov (2007: p. 22) notes that 'the core of dialectics, without which there is no dialectics, is precisely contradiction – the motor and mainspring of development thinking'. In dialectics development normally arises when the actor encounters a situation where habitual thinking and acting is questioned, challenged or inadequate (Virkkunen & Newnham, 2013). The actor then stops, takes a distance and reflects on the contradictory situation. Development happens when the actor creatively overcomes the contradiction through: (a) deepening and broadening his/her understanding; and (b) developing a new and adequate way of acting that changes the system of relationships (*ibid.*). This is achieved by understanding and working on the relationships between two or more related opposites such as master and slave; production and exchange; society and individual, structure and agency. In short, dialectics provide a framework that enables the identification and potential resolution of *contradictions* in and arising from human activity embedded in broad institutional structures and long-term trajectories of change and development (Engeström, 1995). Dialectics' central task is *synchronisation* – a process of revealing, challenging and synthesising competing positions and interpretations by making explicit the implicit assumptions and counter-assumptions (Engeström, 1987). This could be understood as a search for synergy and deeper meaning that addresses the question of practical, functional uniting of opposite elements into a concrete functioning whole.

Developmental work researchers tend to highlight the value of revealing existing contradictions to support learning by expanding: stimulate the development of practices, groups and organisations at the beginning of the interventionist research journey; and during the course of the expansive learning and development journey. This part briefly describes the contradictions that may be identified at the beginning of a DWR journey. An activity always is on certain phase of the cycle and the expansive learning process should identify the phase and the contradictions. The point to remember is that the learning challenges tend to be qualitatively different in different phases of an activity's cycle (Virkkunen & Ahonen, 2004). The following paragraphs describe the different kinds of contradictions.

2.7.1.1 Primary contradiction

When something new and different is introduced into any element of the activity system, tensions are likely to arise between the existing and the new social development. For example, the introduction of a new crop variety in an activity system of farmers will create competition between the old and the new varieties (tools) with regards to allocation of land between them (rules), which may be determined by the motive for growing the new variety (object). It will also demand a new set of skills from among the farmers in terms of how to produce it (tools); and also how to prepare it for food and for sale (object). In some cases it might demand different types and levels of attention from the farming family or community (division of labour). The male and the female farmer in the family (subjects) may have different and competing positions on the new crop, which may be determined by the distribution of costs such as labour or benefits such as food or income. These kinds of tensions that exist within each element are called primary contradiction.

2.7.1.2 Secondary contradiction

If there is a tension between two or more elements of the activity system, a secondary contradiction emerges. For example, when the farmer fails to master or to provide the necessary material support for the production of the new crop variety (subject-tool); or when the agro-ecological conditions are not good for the new crop variety (rules-tool); or when the anticipated demand for it does not materialise and the farmer loses out (subject-object).

2.7.1.3 Tertiary contradiction

A tertiary contradiction is a third layer of tension, which exists between the object of the emerging practice and the old one that it is seeking to replace. It is important to remember that the difference between activity systems is determined by the object – the societal need to seek to address. For example, when a farming community is in the process of changing from a chemical fertiliser and pesticide dependent horticultural production system to an organic production system, it will encounter tensions between maximising production (in the old practice) and paying attention to productivity and ecological sustainability simultaneously (in the new practice). The tensions may manifest themselves through habits, knowledge, experience and relationships of the past and those being learnt and practised. In Sub-Saharan Africa (SSA) and other parts of the world, climate change and HIV/AIDS are some of the major developments that have and are changing the object and motives of agriculture thus creating need for new ways of farming that create tensions with those of the past.

2.7.1.4 Quaternary contradiction

The fourth layer of contradictions is called quaternary and it exists between the new activity system and the others that it interacts with. For example, if farmers have begun organic agriculture, they are likely to encounter obstacles from actors that they relate with. This could be between suppliers (tool-producing activity system) who do not have the right agro-chemicals or extension workers (tool-producing activity system) whose organic knowledge and experience is based on high-input conventional agriculture and cannot offer necessary technical advice. It could also be to do with consumers (subject producing activity system) who are less discerning or unable to pay for premium prices of organic produce. It could also be because the system of inspecting, certifying and labelling organic produce (rule-producing activity system) is underdeveloped or unable to meet demand. It could also be between the conventional and organic farmers operating in the same geographical area.

2.7.2 Identification of contradictions at the beginning of the research journey

At the beginning of a DWR research journey, contradictions are identified through analysing operations, actions and activities for inefficiencies, disturbances, dilemmas, doubts, disagreements, obstacles, clashes and conflicts. They lie underneath experienced discords, conflicts and challenges. The contradictions are identified by the DWR researcher using historical and ethnographic research, and are presented as 'mirror data' for validation by research participants. The developmental work researcher should treat contradictions as potential sources of learning, innovation and development. If resolved, they enable an activity system to become better through expansive learning. However, if not resolved, contradictions can result in the regression of an activity system. The diagram of Figure 2.8 shows the potential sites to find the four types of contradictions.

After identifying the contradictions, the developmental work researcher articulates them and presents them to practitioners for deliberation and validation. The process should lead to the reframing of the challenges so that the practitioners, the developmental work researcher and specialists in the field (who attend a change laboratory workshop) have a shared and agreed understanding of the problematic situation. The next important stage is for them to identify the contradiction whose resolution is most important for the development and expansion of the activity system or practice under review. This takes the process into another layer of further

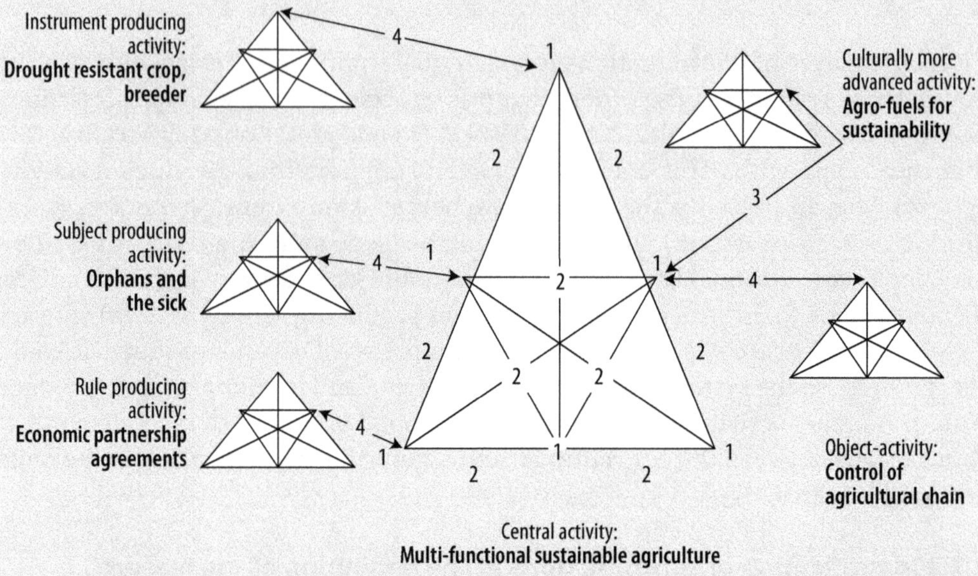

Figure 2.8. Four levels of contradictions and the sites to find them (Mukute, 2010: p. 104).

clarification and negotiation among the workshop participants. When the most important contradiction(s) or structural issue(s) is/are identified and agreed upon, the process of developing the activity system (remediation) begins and attention needs to be paid to contradictions that will arise in the course of implementing the solution in the system.

During the process of remediation, when the activity system is being reviewed and developed, contradictions tend to appear in a particular sequence, from primary to quaternary in that order. This is discussed in the next chapter after introducing the expansive learning cycle in which the emergence of contradictions tends to follow a particular order.

2.8 Agency and practice

The chapter has already pointed out that DWR involves an interpretation of the current situation, an identification of contradictions and the development of solutions that improve and transform practices. Transformative agency is the main focus on this subsection. Put simply, transformative agency refers to individual, relational and collective agency (Engeström *et al.*, 2014).

2.8.1 Individual agency

Agency is human will and capacity to act upon situations towards a purpose (Sibeon, 1999 cited in Lewis, 2002). In order to exercise agency, the individual actor should possess and exercise internal powers and capacities to intervene in events and processes (Barnes, 2000). For this to happen the actor should also 'enter into relationship with surrounding persons, places, meanings and events' Emirbayer (1997: p. 294). These powers, according to Giddens in Barnes (2000), constitute the ability to make a difference by acting on or against external constraints and to transform the structures and systems from which they arise.

2.8.2 Relational agency

Relational agency is concerned with how to relate, cooperate, communicate, work with others and negotiate meaning. It is a capacity associated with *how to be with* others, knowing how to know who, co-create future scenarios, invest in relationships and work together towards a shared purpose (Edwards, 2007; Wals, 2007). Relational agency refers to the social ability to cooperate and communicate with different kinds of people and involves 'the capacity to offer support and to ask for support from others' (Edwards & Mackenzie, 2005: p. 294). Edwards (2007: p. 4) defines relational agency as 'a capacity to align one's thoughts and actions with those of others to interpret one's world and to act on and respond to such interpretation'. It includes capacities to exchange ideas, negotiate, tolerate diversity, and respect differences. It calls for self-knowledge, self-awareness and self-management. It is therefore a very important capacity in multistakeholder innovation processes.

2.8.3 Collective agency

Collective agency is primarily concerned with *how to do with* others, working towards a shared purpose. In the course of improving and changing practice, change may initially be initiated by an individual but a system-wide change can only be achieved through the joint actions of several people or groups. Dean (2009) points out that transformation and emancipation in the globalised and interconnected world can only be achieved through relational-collective undertaking. Collective transformative agency of the people involved in an activity emerges as they understand the historically evolving contradictions behind their daily troubles and generate solution principles that opens the prospects of overcoming them.

2.9 Theoretical and empirical abstraction of practice

Theoretical abstraction forms the central purpose of DWR as Radzikhovskii (1985) notes:

> *Second, the structure of activity has many levels. The structure at the surface, i.e. the external manifestations of the process of activity, is determined by the deep structure of activity; and a person may not even be aware of that deep structure, it may not even be reflected in his self-report. The surface and the deep structure are not identical: the latter is derived from the former through specific transformations.*
>
> *Third, Marxist analysis is a structurogenic analysis. This means in particular, that the basic, generative structures were formed earlier in time than external structures, the structures of the directly observable process. The base structure is a structurally and genetically primary formation, a primary unit relative to the overall process of activity; the entire process unfolds from the structure of this unit. Marxism has provided a model of the method for going from the abstract to the concrete; it has shown how the structuring of the system of social activity is derived from, and unfolds from, the structure of this basic unit. In this sense, the basic unit is also a unit of analysis of the process of activity, i.e. an analysis not at the level of phenomena, but rather a structural genetic analysis.*

Theoretical abstraction leads to theoretical generalisations that are needed for the development of new practices through thought and practical experiments. It builds on a concrete analysis of functional relationships and contradictions between things (objects) (Engeström, 1987). Such relationships of interaction can only exist between entities that are complementary based on a unity of contradictions. A theoretical generalisation often begins as a deviation from previously accepted and codified norms, an exception to the norm as a functioning prototype. The generalisation of a theoretical or design concept is the process of coming into being of a system that 'starts from a new constellation of basic, dynamic relationships of interaction and becomes gradually enriched, expanded and generalized (Ilyenkov, 1977 cited in Virkkunen & Newnham, 2013: p. 45). This process of ascending from the abstract, isolated germ cell of a system to a multifaceted, concrete reality can be understood as the process of coming into being of an improved activity system or practice. As Engeström *et al.* (2014: p. 3) note, this means cultivating 'tentative solutions by means of experimentation, first locally and, when the working solutions are found, by generalising and spreading them through dialogue and further experimentation.'

Empirical generalisation involves comparisons and classification based on judgment-statements (Davydov, 1990). It is concerned with delineating and characterising the object of research based on empirical observations. Empirical generalisations say something that is true about something in all or most cases (Virkkunen & Newnham, 2013). Such generalisations are based on a combination of general attributes, which may be called *classificatory* concepts. For example, in Botany, plants may be classified into families, genera, species, sub-species and varieties based on observations. Empirical analysis enables one to understand and master variation within existing practice. Empirical analysis and generalisations help the development work researcher to develop a good grasp of the history, elements, stakeholders, issues, contradictions and innovations in activity systems and practices.

2.10 Conclusions

This chapter focused on discussing the main concepts that underpin DWR. This was intended to provide the conceptual framing of the resource book, which serves to mediate or connect IAR4D, AIS and People Centred Learning and Innovation on one hand and CHAT-informed DWR methodology on the other. The underlying concept is dialectics, which powers human thinking, action and development through the identification and resolution of contradictions; and the overcoming of obstacles. In doing this, the chapter served as a tertiary tool in Wartofsky's sense. In short, the chapter has provided DWR's argumentative grammar – the logic that guides the use of a methodology and the claims that may arise. It has four tenets namely: '(1) the collective activity systems as unit of analysis; (2) contradictions as the source of change and development; (3) agency as a crucial layer of causality; and (4) transformation of practice as a form of expansive concept formation', Engeström (2011: p. 1). This chapter has built on the broader justification that was made in Chapter 1 and sets the stage for a more specific discussion on how the DWR or formative journey can be designed and implemented in the next chapter – the main 'how' part of this resource manual.

References

Adler, P.S. (2012). The sociological ambivalence of bureaucracy: From Weber via Gouldner to Marx. *Organisation Science, 23(1),* 244-266.

Archer, M. (1995). *Realist social theory: The morphogenetic approach.* Cambridge, UK: Cambridge University Press.

Barnes, B. (2000). *Understanding agency: Social theory and responsible action.* London, UK: Thousand Oaks.

Bartlett, F.C. (1958). *Thinking.* London, UK: George Allen and Unwin.

Bateson, G. (1972). *Steps to an ecology of mind.* San Francisco, CA, USA: Chandler.

Cole, M. (1988). Cross-cultural research in the socio-historical tradition. *Human Development, 31,* 137-151.

Cole, M., & Gagdamaschko, N. (2007). Vygotsky and culture. In: H. Daniels, M. Cole & J.V. Wertsch, (Eds.), *The Cambridge companion to Vygotsky* (pp. 193-211). Cambridge, UK: Cambridge University Press.

Cole, M., & Hatano, G. (2007). Cultural-historical activity theory: Integrating phylogeny, cultural history and ontogenesis in cultural psychology. In: S. Kitayama & D. Cohen, (Eds.), *Handbook of cultural psychology.* New York, NY, USA: Guilford Press.

Daniels, H. (2001). *Vygotsky and pedagogy.* London, UK: Routledge.

Davydov, V.V. (1990). *Types of generalisation in instruction: Logical and psychological problems in the structuring of school curricula.* Reston, VA, USA: National Council of Teachers of Mathematics.

Dean, K. (2009). Agency and dialectics: What critical realism can learn from Althusser's Marxism. In: K. Dean, J. Joseph, J.M. Roberts & C. Wight. *Realism, philosophy and social science.* London, UK: Palgrave and Macmillan.

Del Rio, P., & Alvarez, A. (2007). Inside and outside the zone of proximal development: An eco-functional reading of Vygotsky. In: H. Daniels, (Ed.), *The Cambridge companion to Vygotsky.* London, UK: Routledge.

Donald, M. (2001). *A mind so rare: The evolution of human consciousness.* New York, NY, USA: W.W. Norton.

Edwards, A. (2007). Relational agency in professional practice: A CHAT analysis. *Actio: An International Journal of Human Activity Theory, 1,* 1-17.

Edwards, A., & Mackenzie, L. (2005). Steps towards participation: The social support of learning trajectories. *International Journal of Lifelong Education, 24 (4),* 282-302.

Emirbayer, M. (1997). Manifesto for a relational sociology. *American Journal of Sociology, 103 (2),* 281-317.

Engeström, Y. (1987). *Learning by expanding: An activity theoretical approach to developmental research.* Helsinki, Finland: Orienta-Konsultit.

Engeström, Y. (1995). Objects, contradictions and collaboration in medical cognition: an activity-theoretical perspective. *Artificial Intelligence in Medicine, 7,* 395-412.

Engeström, Y. (2001). Expansive learning at work: Towards an activity theoretical re-conceptualisation. *Journal of Education and Work, 14 (1),* 133-156.

Engeström, Y. (2005). Non scolae sed vitae discimus: Towards overcoming the encapsulation of school learning. In: H. Daniels, (Ed.), *An Introduction to Vygotsky* (2nd ed.) (pp. 157-176). London, UK: Routledge.

Engeström, Y. (2007). Putting Vygotsky to work: The change laboratory as an application of double stimulation. In: H. Daniels, M., Cole & J.V. Wertsch, (Eds.), *The Cambridge Companion to Vygotsky* (pp. 363-382). Cambridge, UK: Cambridge University Press.

Engeström, Y. (2008). *From design experiments to formative interventions.* Paper presented at the ISCAR Conference. September 9-13, 2008. San Diego, CA: University of California.

Engeström, Y. (2011). *From design experiments to formative interventions.* Helsinki, Finland: CRADLE, University of Helsinki.

Engeström Y., & Sannino, A. (2011). *Volition and agency in organisations: An activity theoretical perspective.* Paper presented at the Conference of Association of International de Management Stratégique (AIMS). June 9, 2011. Nantes, France: AIMS.

Engeström, Y., Sannino, A., & Virkkunen, J. (2014). On methodological demands of formative interventions. *Mind, Culture and Activity, 21,* 1-11.

Eskola, A. (1999). Laws, logics and human activity. In: Y. Engeström, R., Miettinen & R.L. Punamäki, (Eds.), *Perspectives on activity theory.* Cambridge, UK: Cambridge University Press.

Haapasaari, A., Engeström, Y., & Kerosuo, H. (2014). The emergence of learners' transformative agency in a change laboratory intervention. *Journal of Education and work.* Retrieved from: http://dx.doi.org/10.1080/13639080.2014.900168.

Hegel, G.W.F. (1969). *The logic of Hegel.* Oxford, UK: Clarendon Press.

Hill, R., Capper, P., Wilson, K., Whatman, R., & Wong, K. (2007). Workplace learning in the New Zealand apple industry network: a new co-design method for government 'practice making'. *Journal for Workplace Learning, 19 (6),* 359-376.

Ilyenkov, E.V. (2007). Our schools must teach how to think. *Journal of Russian and East European Psychology, 45 (4),* 9-49.

Leont'ev, A.N. (1978). *Activity, consciousness and personality.* Eaglewood Cliffs, NJ, USA: Prentice-Hall.

Leont'ev, A.N. (1981). *Problems of the development of the mind.* Moscow, Russia: Progress.

Lewis, P.A. (2002). Agency, structure and causality in political science: A comment on Sibeon. *Politics, 22 (1),* 17-23.

Lizares-Bodegon, S., Gonsalves, J., Killough, S., Waters-Bayer, A., Van Veldhuizen, L., & Espineli, M. (2002). *Participatory technology development for agricultural improvement: Challenges for institutional integration.* Cavite, Philippines: International Institute of Rural Reconstruction (IIRR).

Martinez-Alier, J. (2002). *The Environmentalism of the poor: A study of ecological conflicts and valuations.* Cheltenham, UK: Edward Elgar.

McMichael, P. (2014). Historicising food sovereignty. *Journal of Peasant Studies, 41 (6),* 933-957.

Mukute, M. (2010). *Exploring and expanding farmer learning in sustainable agriculture workplaces.* Unpublished PhD thesis. Grahamstown, South Africa, Rhodes University.

Mukute, M. (2014). *Phiri award for farm and food innovators: Report on cases studies of nominated farmer innovators* (unpublished). Harare, Zimbabwe: Phiri Award for Farm and Food Innovators Trust.

Mukute, M., & Lotz-Sisitka, H. (2012). Working with cultural historical activity theory and critical realism to investigate and expand farmer learning in southern Africa. *Mind, Culture and Activity, 19 (4),* 342-367.

Osano, P.M., & Adam, R.I. (2014). An intergenerational perspective towards increasing young people's contribution to agriculture in Sub-Saharan Africa: Agriculture and farming in Sub-Saharan Africa – from challenges to opportunities. In: P.B. Corcoran, B.P. Hollingshead, H. Lotz-Sisitka, A.E.J. Wals & J.P. Weakland, (Eds.), *Intergenerational learning and transformative leadership for sustainable futures.* Wageningen, the Netherlands: Wageningen Academic Publishers.

Peal, D., & Wilson, B. (2001). *Activity Theory and Web-Based Training.* Available at: http://www.ceo.cudenver.edu/~brent_wilson/acttheory.html.

Radziknovskii, L.A. (1985). Activity: Structure, genesis and unit of analysis. *Soviet Psychology 32 (2),* 35-53.

Sawchuk, P.H. (2003). *Adult learning and technology in working class life.* Cambridge, UK: Cambridge University Press.

Sayer, A. (2000). *Realism and social science.* London, UK: Sage.

Simovska, V. (2005). Participation and learning about health. In: S. Clift & B.B. Jensen, (Eds.), *Health promoting school: International advances in theory, evaluation and practice* (pp. 173-192). Copenhagen, Denmark: Danish University of Education Press.

Simovska, V. (2008). Learning in and as participation: A case study from health promoting schools. In: A. Reid, B.B. Jensen, J. Nikel & V. Simovska, (Eds.), *Participation and Learning: Perspectives on Education and the Environment, Health and Sustainability* (pp. 61-80). Bath, UK: Springer.

Tolman, C. (1981). The metaphysics of relations in Klaus Riegel's 'Dialectics' of human development. *Human Development, 24,* 33-51.

Tomasello, M. (1999). *The cultural origins of human cognition.* Cambridge, UK: Harvard University Press.

Tomasello, M., Kruger, A.C., & Ratner, H.H. (1993). Cultural learning. *Behavioural and Brain Series, 16,* 495-552.

Vadeboncoeur, J.A., & Collie, R.J. (2013). Locating social and emotional learning in schooled environments: A Vygotskian perspective on learning as unified. *Mind, Culture and Activity, 20 (3),* 201-225.

Virkkunen J., & Schaupp, M. (2008). *From change to development: Expanding the concept of intervention.* Paper presented at the ISCAR Conference. September 9-13, 2008. San Diego, CA: University of California.

Virkkunen, J., & Ahonen, H. (2004). Transforming learning and knowledge creation on the shop floor. *International Journal of Human Resources Development and Management, 4 (11),* 57-72.

Virkkunen, J., & Newnham, D.S. (2013). *The change laboratory: A tool for collaborative development of work and education.* Rotterdam, the Netherlands: Sense Publishers.

Virkkunen, J., Engeström, Y., Helle, M., Pihlaja, J., & Poikela, R. (1997). Change laboratory: A tools for transforming work. In: T. Alasoini, M. Kyllonen & K.A. Kasvio, (Eds.), *Workplace innovations: A way of promoting competitiveness, welfare and employment* (pp. 157-174). Helsinki, Finland: National Workplace Development Programme.

Visvanathan, S. (2006). Alternative science. *Theory, Culture and Society, 23,* 164-169.

Vygotsky, L.S. (1934). *Thinking and speech.* Chapter 1. Available at: http://www.marxists.org/archive/vygotsky/works/words/ch01.htm.

Vygotsky, L.S. (1978). *Mind in society: the development of higher psychological processes.* Cambridge, UK: Harvard University Press.

Wals, E.J. (2007). Learning in a changing world and changing in a learning world: Reflexively fumbling towards sustainability. *Southern African Journal of Environmental Education, 24,* 34-62.

Wartofsky, M.W. (1979). Perception, representation, and forms of action: Towards an historical epistemology. In: M.W. Wartofsky, (Ed.), *Representation and scientific understanding, Boston studies in the Philosophy of Science, Volume 48*. Dordrecht, the Netherlands: Reidel.

Chapter 3. Illuminating the DWR methodology

3.1 Introduction

The previous chapter discussed the roots of DWR this chapter focuses on *how* DWR can be used to enhance the expansion of developing systems through epistemic or learning actions. These are carried out in an expansive learning process that identifies and resolves contradictions that emerge and evolve in an activity or activity system. Put differently, an activity is a practice and an activity system is a group of people involved in a practice. The group could be as small as a local farmer group or as large as an international farmer, civil society, government, research or private organisation.

DWR is based on the understanding that practices and systems are always developing and in motion as suggested by the ontology of developing systems. For example, in agriculture, farmers and scientists are always looking for ways to improve their practices, sometimes by addressing contradictions. But the point of DWR and therefore of this chapter is to articulate a process by which agricultural stakeholders can periodically and strategically learn and innovate together and overcome the main contradictions that they encounter in agriculture. The process helps to speed up the development of practices and systems.

Through the methodology, researchers, practitioners and specialists jointly research, learn and act to improve practices and relationships. Meaningful interaction between DWR researchers and participants only takes place when they share an object in which they have a common or reciprocal concern (Miettinen, 2004). Raelin (2007) notes that DWR is underpinned by an *epistemology of practice*, which encourages knowing-in-action, human agency and social change.

The spine of the methodology, which is discussed in this chapter, is the CL workshop method. CL workshops can be used to support IPs in agriculture and equivalent structures and processes in NRM. It is DWR's boundary-crossing or horizontal expansion approach and tools that enable people from different backgrounds to learn and work jointly to improve practices or create new forms of practice (Kerosuo & Engeström, 2003). The purpose of this chapter is to provide guidelines on how to work with DWR and the CL method in the expansive learning process.

3.2 DWR in short

Here is how one of the leading scholars on CHAT and DWR methodology described it:

> *DWR is the study of realistic possibilities. This implies a sequence of steps. The first step is to identify a general, historically involved inner contradiction in a type of activity [practice]. The contradiction specifies the tasks of further learning and area of possibilities. The second step is to find a concrete, local activity, within which the contradiction manifests itself in a specific aggravated form and in which there exists motivation and resources to search for a concrete solution and principle of overcoming the contradiction that in the first instance, shows a specific possibility. The process can also begin in an instance where there is an acute need for finding a solution in a local activity. The third step is the concretisation and generalisation of the found principle. The research results of a study can comprise these steps. The analysis and documentation of contradictions is an important result, even if a solution principle has not been achieved and a solution principle is important, even if it has not yet been generalised within a specific study. The peculiar aspect of DWR is the idea, that a possibility is there only if there is a collective actor who has taken its realisation as its object, so transformative agency is not a general phenomenon but an aspect of a historically specific developmental possibility, although its manifestations can be empirically studied also formally as features of behaviour ... A central idea that is behind expansive learning and DWR but has seldom been explicated in their presentation is Marx's idea of the progressive socialization of forces of production that implies that there is a general tendency for broader and tighter division of labour in production and the replacement of local idiosyncratic knowledge with general, scientific knowledge [mature concepts]. That means that the solution for an inner contradiction within a local activity is in the widening of collaboration, specialization and division of labour.*
>
> *(Virkkunen, personal communication)*

3.3 Acknowledging on-going innovations

DWR seeks to enhance local innovation by identifying and drawing on them to overcome the structural and more deep-seated challenges being faced by people. As such it is important to acknowledge the innovations that may be taking place in an area and activity of study. The story below shows how an innovative farmer in Zimbabwe went about improving an agricultural practice in his agro-ecological setting, and shared the improved practice with some of the farmers in his community (Mukute, 2014). The story is one of many that were documented for the Phiri Award for Farm and Food Innovators Trust that was founded in 2013 for the purpose of recognising and encouraging farmer innovation.

> *Sixty-eight year old Bouwas Mawara has been farming in low-rainfall Mazvihwa communal area of Zimbabwe since 1976. But his goal of producing crops in the dry and wet seasons was initially undermined by water shortages and dry spells. In 1980 he questioned the use of the officially-promoted contour ridges that were designed to drain away water from the fields, which were often short of water. It was then that he designed contours that trap and retain water and allowed it to sink into the ground, thus recharging the water table. He says 'Water means to soil's life what blood means to a person's life.' When this worked, he developed a complex system of clay pipes that distributed water in his plot. After addressing the water problem, he worked on improving soil fertility, which he had identified as the next obstacle to producing enough food for household consumption and for sale. By 1986 he had so much water in his deep contours that he started keeping three kinds of fish thus improving the nutrition security for his household. He also sold some of the fish for income. Over the years, he has developed a cropping system that enables him to plant three crops annually, thus increasing production from the same piece of land. Along his innovation journey, he encountered some problems. For example, some farmers questioned his sanity when he was digging deep contours that were different from the standard ones. Others stole crops and fish but he soldiered on. But there were others who came to seek his advice so that they could also benefit from his innovations. His subsequent and strategic response was to form a local farmer innovators' group called Hupenyu Ivhu (Life comes from the soil) in 1989, nine years after beginning his innovation journey. Since then, he has used the group to spread and improve the set of innovations among farmers living in Mazvihwa. The initial result was improved food security and livelihoods for his family. The adoption of the improved practice by at least 160 farmers in his neighbourhood has improved the community's well-being.*

In addition to documenting the stories of individual farmer innovators, the study analysed the patterns of local innovation processes (Mukute, 2014) and the result of this is captured in Textbox 3.1. The patterns show some similarities and differences between DWR and the local innovation processes. The point is not to replace one approach with the other but to illustrate how DWR can build on the innovativeness

Textbox 3.1. Trajectories of farmer innovation processes documented in Zimbabwe.

Twelve (12) farmers were nominated as innovation farmers under the maiden Phiri Award in 2014 in Zimbabwe. The stories that were documented revealed the following pattern of farmer innovation, which is not linear but iterative:

a. *Problem/opportunity identification.* Each innovator responded to an adversity or limitation in agricultural production that he was facing. For example, one farmer innovator responded to the washing away of top soil by Cyclone Eline. Occasionally, the starting point was the mere questioning of prevailing wisdom concerning one or more aspects of farming. For example, the entry point of another innovative farmer was: do contour ridges that are promoted by agricultural authorities and that drain away water from our fields serve our crop production interests?
b. *Initial experimentation, trial and error.* The second stage may be called trial and error experimentation. During this phase the innovator draws on different knowledge sources and material resources to tackle the challenge: traditional knowledge, formal technical knowledge, own experiences and observation, and fellow farmer cognition. But this is also a period during which farmers are discouraged from trying out new things.
c. *Initial breakthrough.* At this stage, the creative idea works on the ground. The innovator becomes confident and begins to reap benefits from his/her effort. Criticism gradually dies down. Some members of the community begin picking some aspects of the solution and also begin to directly or indirectly benefit from the innovation. However, others move from discouraging to sabotaging the application of the innovation as a way of proving their point.
d. *Perfection and expansion of innovation.* When the innovative solution is being applied, new challenges arise and the innovator tends to keep on looking for additional creative ways to build on the innovation. For example, those whose innovations initially focused on soil or water improvement expanded into agricultural production and productivity. Improvements at this level are based on experience from the practical application of the innovative idea. Such application is done by people with different backgrounds and resources, working in agro-ecological settings that are slightly different. The experience that is used to perfect or expand the innovation is gathered over several seasons. Personal experience and feedback from early adopters and other interested stakeholders assist the innovator to consolidate the innovation. The process of developing and perfecting the innovation generally takes more than one agricultural season as certain realities have to be taken into consideration and do not always manifest themselves in one year. For example, droughts and above average rainfall may occur five years apart and may need to be taken into account.
e. *Scaling out the innovation.* When there is sufficient evidence that the innovation works, the innovator responds to demand by creating a structure that enables the dissemination of the innovation. This may arise from the 'inefficiencies' of having to attend to many fellow farmers who come to enquire at different times and thus 'disrupting' the innovator from performing his farming activities. In other cases scaling out takes place when local development organisations recognise the innovation and find ways of linking other farmers with the innovator. Most farmer innovators discussed in this report experienced difficulties and struggles in the scaling out process. (Mukute, 2014)

inherent in agricultural stakeholders, starting with but going beyond the farmers as innovators.

In India, innovative practice that is based on building on indigenous and traditional knowledge is much more robust. Some scholars such as Professor Anil Gupta of Honey Bee have been at the forefront of documenting and stimulating local innovation (Mukute, 2006; Radjou *et al.*, 2012). One of the most interesting stories that show local ingenuity and innovativeness was triggered by a crisis arising from a 2001 earthquake:

> *A potter by trade, Prajapati had for years been experimenting with clay to produce a variety of durable goods ... Prajapati explained how this clay fridge – the Mitticool ... works: The fridge consumes no electricity, is 100% biodegradable, and produces zero waste during its lifetime ... But this inventor and his personal story are even more impressive ... he didn't even finish high school. In 2001, an earthquake had devastated Prajapati's village and the surrounding area. Reading a report of the devastation in the local newspaper, he noticed a photo caption: 'Poor man's fridge broken!' The photo featured a smashed earthen pot commonly used by villagers to fetch water and keep it cool. And though the newspaper had called it a fridge in jest, it triggered Prajapati's first eureka moment. 'Why not use clay,' he thought, 'to make a real fridge for villagers – one that looks like a typical fridge, but is more affordable and doesn't need electricity?' ... The positive health and lifestyle benefits of owning a fridge in the desert village where fruit, vegetables, and dairy are available only intermittently would be tremendous. He experimented for several months and eventually had a viable version of Mitticool that he began selling to people in this own village. The fridge – which costs around US$50 – was a hit. Prajapati worked tirelessly on design improvements, and began selling Mitticool across India, and then internationally. He couldn't keep up with the rising demand ... Then he had a second eureka moment. Why not transform pottery from an artisanal craft to an industrial process? He leveraged his traditional knowledge of pottery to mass-produce goods that met modern consumer needs. Prajapati first developed an entirely new and more efficient method of working with clay. Then he began training women in his village in these industrial pottery techniques and finally hired them to work in his new factory ... He soon built other products from clay, such as a nonstick frying pan that retains heat longer than other frying pans and costs a mere US$2. From one man and one idea has grown a frugal yet fruitful industry, one that employs large numbers of people in his own community and serves consumers in India and abroad.*
>
> *(Radjou* et al.*, 2012: pp. 1-3)*

The spirit of DWR is to create spaces where practitioners such as farmers and scientists such as agronomists, breeders and sociologists collectively analyse agricultural obstacles, learn and innovate to improve agricultural practices and systems. They do this with the assistance of a DWR process specialist whom we also call a developmental work researcher, who in turn is guided by the theoretical concepts underpinning DWR and discussed in the previous chapter. The developmental work researcher is further guided by DWR principles and associated CL method, which are discussed in this chapter.

3.4 Noting of similarities between people-centred learning and innovation and DWR

DWR should be anchored on on-going agricultural research and development processes that have a similar intention and conceptual basis and to which it can add value. Some of these have already been discussed in Chapter 1. People-centred learning and innovation is another one and its similarities with DWR are listed below (Table 3.1). The linkages are intended to help someone familiar with People Centred Learning and Innovation to be in a better position to work with DWR.

3.5 DWR and the expansive learning process

3.5.1 The essence of DWR

DWR was initially developed in Finland in the early 1980s (Engeström, 1987) to overcome the challenges of a historical and individual-centred workplace learning. It emerged from the need for a research methodology that deals with both the practical and theoretical dimensions of problematic situations that people encounter in productive work. The two main intended outcomes are the accumulation of scientific knowledge and the development of practices or activities in which the methodology is applied. DWR is a formative research methodology that seeks to reveal and address the needs and possibilities for development in a practice or a community. It involves *transforming experiments* that radically restructure the environment, established institutional forms and values, and activates previously unrealised human potentials (Bronfenbrenners, 1977). DWR contributes to practice and theory through:

Table 3.1. Similarities between People Centred Learning and Innovation and DWR.

Dimension	People-centred learning and innovation (Scoones, Thompson & Chambers, 2008)	CHAT-informed DWR
Mental model of activities	Innovation network centred on co-development; involving multi-stakeholder processes and messy partnerships	DWR involves joint learning, innovation and agency development among subjects of multiple interacting activity systems that have a shared object
Farmers seen by scientists as	Partners, collaborators, entrepreneurs, innovators; organised group setting the agenda, 'the boss'	One of the major object-producing activity systems through active participation in (research) agenda setting processes. Farmers are also viewed as co-participants in the modelling, examination and strengthening of cross-learning, innovation and agency development
Scientists as seen by farmers	One of many sources of ideas	One of the tool-producing (technical and conceptual) activity systems to learn with and from, as well as to influence, innovate and act with
Knowledge and disciplines	Extra/trans-disciplinary – holistic, multiple culturally rooted practices	Societal and cultural development that draws on the distributed cognition of actors with different backgrounds and professions as a basis for theoretical generalisations
Farmers' roles	Empowered, co-generators of knowledge and innovation; negotiators	Co-creators of knowledge and innovations, whose motives also determines the direction of the joint research and innovation efforts
Scope	Beyond the farm gate – multi-functional agriculture, livelihood, food systems and value chains across multiple scales from global to local; long time frames	The agricultural activity systems and those that it interacts with
Drivers	Responsiveness to changing contexts: markets, globalization, climate change, organised farers, power and politics	Problematic situations and contradictions within and between activity systems; jointly constructed Zone of Proximal Development (ZPD) that is reached through overcoming problematic situations, double-binds or historically accumulating structural contradictions
Intended outcome	Capacities to innovate, learn and change	Future-oriented general solution such as transformative agency; collective learning; conceptual and material tools, and theoretical generalisation
Institutions and politics	Central dimensions of change	Interacting activity systems comprise the institutional make-up while transformative agency on structures and systems has political implications
Innovators	Multiple actors, learning alliances	Subjects from the interacting activity systems, which include practitioners and specialists

- *Learning and production of knowledge.* Generating, articulating and testing design principles, which are oriented towards the production of new knowledge that can be theoretically generalised. The outcomes of learning itself include expanded objects, new collective practices, including the practices of thinking and discourse (Engeström & Kerosuo, 2007). Lindley and Lotz-Sisitka (2014), drawing on several CHAT scholars, identify the following learning actions as being enabled by the DWR process:
  - sense and meaning making by participants through understanding their work in context;
  - a broader orientation, perception and understanding of the activity or practice under review;
  - development of new knowledge and creation of creative work practices, including practices of thinking and engagement;
  - creation of new professional practices that cross traditional professional boundaries; and
  - re-conceptualisation of problematic situations and the development of tools to overcome them, leading to the expansion of the activity.
- *Improvement of practice and activity systems.* Producing innovative solutions for improving practices so that they become qualitatively different or for the development of new forms of activity/practice. This is achieved through drawing on the distributed cognition that exists among interacting stakeholders or research participants. The innovations become the means by which activities get transformed through application, refinement and integration. They culminate in the development of the developing activity system, which entails the overcoming of internal contradictions of a practice as well as the external ones that may constrain it.

DWR is a step-wise process, which is summarised by the expansive learning process. The expansive learning process, and therefore the DWR, seeks to:

- Overcome historically accumulating contradictions. In agriculture this means exploiting the existing conflicts and dissatisfactions among farmers, development practitioners, researchers and others involved in and affected by the agricultural practices.
- Draw on the strengths of joint analysis and collaborative learning to develop innovative solutions to current limitations.
- Bring about concrete transformation of current practice and existing relationships.

3.5.2 Expansive learning

> *In expansive learning and development processes, there is a dialectic between objective material changes and changes in the actors' understanding and agency. The historical and actual-empirical analyses anchor the actors' thinking and search for a new solution to the objective situation in the activity, which they can change through their practical action. In this process the changes in the activity system create inner contradictions within and between its elements and in relation to other activities ... The location and nature of the inner contradictions change as the transformation proceeds creating challenges for further learning and creation of expansive solutions.*
>
> *(Virkkunen & Newnham, 2013: pp. 51-52)*

The process of expansive learning entails collaborative learning and seeks to address new and emerging problems, creating new knowledge and building local resilience. It offers a framework for understanding forms of learning that do not adhere to standard models of mastery. It is concerned with knowledge creation and application iteratively. Learning becomes expansive when it contributes to an enlarged room to manoeuvre for the individual whereby new learning possibilities are formed. Individual and collective learning takes place by going beyond the boundaries of individual subjectivity through immediate cooperation towards the realisation of common interests of collective self-determination against dominant partial interests. This makes expansive learning an inherently multi-voiced process that involves debate, negotiation and careful organisation and coordination. One of the critical aspects of expansive learning is its reliance on self-organisation from below, which manifests itself in the 'creation of networks of learning' that transcend institutional boundaries. Expansive learning is also concerned with material *transformation of vital relations* through individual, relational and collective agency development.

A full cycle of expansive learning usually takes place over two or more years and comprises several interconnected smaller or miniature cycles of learning (Bodrožić, 2008; Engeström *et al.*, 2007). But if the discontinuities that arise are not bridged by further learning and innovation, the full expansive learning cycle will not be completed. For example, Lindley (2014: p. 328) concludes:

> *The PhD research project could be seen as one of the smaller expansive learning cycles, which has ended with mundane discontinuities ... These discontinuities ... now need to be bridged by management to enable the beginning of the next cycle, to ensure the continuity of the expansive learning process and catalysing*

> *change as part of the larger expansive learning cycle. This will further the organisational learning and development of Mondi in a progressive never ending cycle of continuous improvement and change, and endless smaller cycles of expansive learning where the end of one cycle becomes the beginning of the next.*

The stages of development in an expansive learning process, which are iterative, are illustrated in Figure 3.1. Expansive learning derives depth from iterative processes of action, reflection and modelling in one expansive cycle (Brall, 2007: p. 92). The arrows point in both directs because the process is iterative. There is usually need to go back and change things because new insights have been generated.

Expansive learning stages can be summarised as follows:

1. *Questioning.* The developmental work researcher uses ethnographic research to question practice focusing on *concrete experiences and actions* related to the practice largely based on concrete experiences of the concerned people. This is augmented by literature review and document analysis. The developmental work researcher should spend some time with the people concerned, appreciating their realities. The data that is generated at this stage is compiled in the form of 'mirror data' and captures the critical incidents, main challenges and tensions that are being experienced in the community, usually in a practice. It is analysed using second or third generation activity system as a unit of analysis.

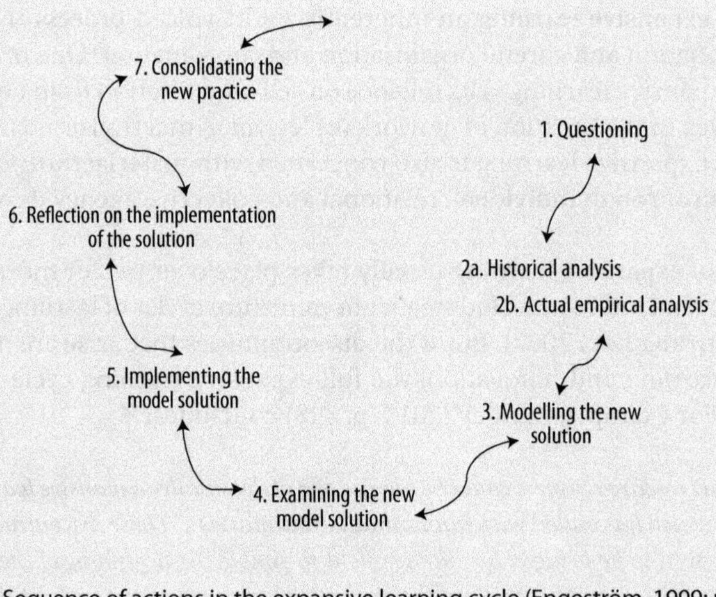

Figure 3.1. Sequence of actions in the expansive learning cycle (Engeström, 1999: p. 384).

2. *Analysing.* The second stage of historical and actual-empirical analysis examines the history of the practice or activity system under review and the broader societal activity in which it is embedded. The emphasis shifts from the current vague feelings and tensions to historical changes and emerging tensions of the developing activity system or practice. The examination should also examine the historically formulated relationships such as those between farmers, extension workers, scientists and policy makers. The developmental work researcher presents the mirror data in a CL workshop that is attended by practitioners whose practice is under review and specialists that have a stake in the practice. The CL workshop provides feedback on the mirror data and further analyses the practice and activity systems in a CL workshop. This involves asking 'why' questions to jointly analyse historical data concerning a practice and associated activity systems. This constitutes the historical-genetic analysis, which should produce causal explanations by tracing the origin and evolution of the disturbances or obstacles in the practice. The workshop then conducts an empirical analysis of the problematic situation focusing on the inner systemic relations of the activity systems. This enables the identification linkages between individual actions and the collective activity system through linking everyday problems and tensions to fundamental contradictions. The products of the analysis are structural contradictions that become the driving force for joint learning, innovation and agency development. These occur within and between elements of the activity system under review.
3. *Modelling.* The research participants, with some assistance from the developmental work researcher, collectively develop a model solution to the structural contradictions. The developmental work researcher's main role is to introduce a conceptual tool that would be used to understand the problematic situation or contradiction better as a basis for developing an innovative solution. The tool is called a second stimulus while the problematic situation or contradiction is called the first stimulus. Participants can and often eventually decide on the second stimulus to work with. The participants develop a model solution based on an expanded understanding of their activity system(s) and associated contradictions. Another key role of the developmental work researcher at this stage is to create an environment that is conducive to interaction, dialogue and inter-subjectivity. Individual actors should be prepared to see the link between their roles and actions on one hand and the interrelated actions and collective activity on the others (Engeström, 1987). Participants co-create new ways of working or engaging with practice in order to tackle the contradictions and challenges that are being faced by a group of people (subjects/community) collectively involved in a practice. This is achieved through tapping into the

distributed cognition of research participants. The model solution represents the *initial breakthrough or the potential germ cell* (Davydov, 1990).

4. *Examining the model solution.* Research participants test the model solution during CL workshop in order to fully grasp its dynamics, potentials and limitations and to improve it as necessary. The developmental work researcher facilitates the process of examining the model solution that participants develop and present. Examining the model solution is an attempt at making a link between the model solution and what is happening on the ground – the concrete situation. It may be regarded as the first layer of reflection on the model solution. This process should adjust and enrich the potential germ cell, which is a relation, action, concept or artefact, which contains properties that will stimulate the further development (of the practice) and interaction with other processes. The germ cell helps the people concerned to reconstruct the whole as a systematic whole that exhibits the essential features of the unit that formed the basis of the intervention (Blunden, undated).
5. *Implementing the model solution.* Participants implement the model solution in their workplaces in the absence of the developmental work researcher. Implementing the model solution entails working with the model in real life situations and monitoring its impacts outside CL workshops. Different kinds of *resistance* are encountered in implementing the model solution in a real life situation, potentially creating tensions that should be dealt with. For example, the introduction of a new conceptual tool, is likely to cause tensions between the new tool and the old, and then between the tools and other elements of the activity system. This creates primary and secondary contradictions respectively. Later as the new activity begins to evolve, some tensions are likely to arise between the object and motive of the emerging activity system and that of the old the activity system, causing tertiary contradictions. The process of integrating the improved or new practice into societal life causes quaternary contradictions that should be resolved before the practice is consolidated. In present day agriculture climate change adaptation and mitigation, ecological sustainability and social justice have become critical objects and motives for agricultural practice.
6. *Reflecting.* The participants and developmental work researcher meet in a CL workshop sometime after the implementation of the model solution to reflect on the implementation process and emerging outcomes. They jointly assess and refine the model solution based on the experiences of working with it in real life situations. This should result in the internal completeness and *stabilisation* of the solution in the practice and participating activity systems. At this stage the tertiary contradictions get resolved.

7. *Consolidation*. Participants incorporate the refined model solution as part of a practice while at the same time influencing neighbouring or interacting activity systems to transform through integrating it resulting in the stabilisation of the practice. This is what Tomasello (1999) calls the ratchet effect. It enables the new practice to survive and thrive. For example, associated policy and strategic changes may be made at activity level. This should result in the *generalising* of the new or modified practice. In AIS, IAR4D and People Centred Learning and Innovation language, this entails the scaling out and up of an innovation. Such generalisation means that the quaternary contradictions that arise between the activity system or the practice and those that it interacts with get resolved. The innovation then becomes part and parcel of the cultural tools available in the community and society. This level of transformation of a practice is reached when the solution becomes anchored in the social and emotional everyday actions leading to societal development. The search for new and emerging contradictions is continuous and does not stop when an innovation has been adopted. This way, society continuously improves and the activity systems become *more advanced.*

3.5.3 DWR and metaphors

Cultural evolution and learning scholars, such as Donald and Tomasello; CHAT theorists, such as Vygotsky and Engeström; and developmental work researchers, such as Seppänen, Koskimies and Mukute, have drawn on, utilised and looked for metaphors because they capture complexity effectively. For example, Tomasello (1999) uses the metaphor of the ratchet effect to refer to the faithful transmission of culture, which provides a platform for further innovations. Vygotsky (1934) likens 'word' to a 'living cell' in looking for the basic unit of analysis associated with verbal thinking and coined the term Zone of Proximal Development, which was later to be elaborated on by Engeström (*ibid.*) using metaphor. He (*ibid.*: pp. 12-13) notes:

> *In the tradition of activity theory, a key metaphor for development is that of a zone. Often the zone of proximal development is interpreted as a vertical step which leads to a higher stage or level. I find it more useful to think of it as a terrain of an activity to be dwelled in and explored, not just a stage to be achieved or a space to be crossed. The zone is explored by moving in it ... The dwellers create trails and the intersecting trails gradually lead to an increased capability to move in the zone effectively, independently of the particular location or destinations of their subjects. However, the zone is never an empty space to begin with. More than that, it has pre-existing dominant*

> *trails, landmarks and boundaries made by others, often with heavy histories and power invested in them ... When new dwellers enter the zone, they both adapt to the dominant trails and struggle to break away from them. The latter leads to critical conflicts and double binds.*
>
> *(Engeström, undated, pp. 12-13)*

Seppänen & Koskimies (2002) capture the long-term interests of sustainable agriculture as 'farming across years'. Mukute (2010a) uses the metaphor of streams of agentive talk to denote the connections between speech turns and discovers that practitioners use metaphors to communicate different kinds of contradictions, agency and reflexivity. For example, one practitioner noted that youths avoided agriculture because it was neither 'cool' nor 'sexy'. The organic industry in South Africa was likened to a baby that needed time and effort to grow and mature as one of the CL workshop participants put it:

> *Basically we have got a baby; the organic industry is a baby, and we are expecting it to be a varsity professor. That is not going to happen. It has got to start from grade 1 and it is going to have its experiences. It is going to have its cuts and bruises and this is what we are going through now.*

In Lesotho the need to build relational and collective agency between two Machobane Farming System promoting organisations put figuratively as follows:

> *Me' is actually emphasizing on the importance of collaboration between the two top officers [in the organisations]. She has realised that the gap was small at the beginning but as time went by she realised that the gap that is between the two officers was growing, until now that there is a huge gap ... It seems there's a donga in between the two offices, which needs to be closed so that we can improve the MFS.*

Metaphors are concerned with the use of one idea to express another by making an implicit comparison, often with the aid of images. They are cognitive resources that help adults to construe whole events or situations in terms of another such as love as a journey and anger as heat (Lakoff, 1987). Metaphors are used in everyday language to link the familiar and the not so familiar or to describe complex phenomena – and allow people to communicate complexity that literal discourse alone cannot (Gibbs, 1994). Metaphors are also figurative ways of talking to children by drawing analogies between the concrete domains of what they know from their sensory motor experiences and the more abstract domains of adult interaction and social and mental life (Tomasello, 1999). Danermark *et al.* (2002) note that metaphor is an abductive mode of inference which assists people to make meaning of complex discourses. Conceptual metaphors are also used in theories to link the abstract to the concrete and to help people understand theories and concepts. Metaphor therefore shapes communication, and may condition the way people think and act (Tomasello, 1999).

It is therefore important to pay attention to metaphors that may be made during the generation of mirror data. At the same time, the developmental work researcher should look for metaphors that express the complex theoretical models and concepts that he/she has to work with.

3.5.4 Sequence of contradictions in the course of expanding a practice

The sequence of contradictions in a developing activity system or practice is closely associated with but different from the expansive learning action. This is because the focus here shifts to the changes taking place in the practice and not in the individuals causing the changes. Virkkunen & Newnham (2013: p. 53), provide a succinct explanation of the temporal sequence of contradictions along the journey of expanding or transforming an activity system or practice:

> *The sequence of contradictions ... can be understood as a temporal sequence in the transformation of an activity system. At the same time it represents the process of descending from the immediately observable surface of the activity to the core of its inner structure and ascending from an abstract new principle to a new concrete form of the activity. In actions of questioning the primary contradiction appear as increasingly troubling but diffuse tensions and disturbances. In actions of analysis, manifest and aggravated secondary contradictions can be identified between elements of the system. In actions of implementation, tertiary contradictions appear between the new model of the activity and the existing one. In actions of consolidating and generalising, quaternary contradictions are encountered between the changed activity and the neighbouring activities.*

When a model solution is formulated and implemented, it creates tensions within the activity system first; and subsequently between the new and the old activity system; and finally between the new activity system and the neighbouring activity system. The types and sequence of contradictions is discussed below:

1. *Primary contradictions.* They are identified through the questioning of standard practice to establish the *need state* of a practice or activity system in a CL workshop at the beginning of a research journey. Primary contradictions may express themselves through vague discontentment that is directed at other people. 'The expansive learning cycle begins with a need state characterised by vague feelings of rupture and tension as well as differing views about what does not work and practical experiences of it' (Kontinen, 2013: pp. 121-122). The tension and ruptures often carry with them emotions of the people who may blame others or defend themselves. The disturbances that are important and/or recurring are then collectively identified, noted and carried forward for analysis.
2. *Secondary contradiction.* A secondary contradiction in this case comes into view when research participants and the developmental work researcher conduct a historical and empirical analysis to reveal the systemic causes of the problematic situation (Virkkunen & Newnham, 2013) and find the causal explanations that give rise to discontentment and tensions. This helps them to contextualise the problematic situation. The analysis reveals *double binds* in the structural features of the activity system, which show up in the relationships between elements of the activity system. The English metaphor that is used to express a double bind is 'between a rock and a hard place'. The research participant experiences double binds in the form of dilemmas, motive conflicts and inner doubts, which may paralyse action (Engeström & Sannino, 2011). The identification of systemic causes urges research participants to shift their attention from people to the goals of the actions in the practice or activity system. Secondary contradictions are overcome through an expansive reconceptualisation of the object and motive of the activity and the creation of corresponding new elements of the structure of the activity system (Virkkunen & Newnham, 2013).
3. *Tertiary contradiction.* A tertiary contradiction emerges during the implementation of a solution to address the secondary contradictions in the course of improving a practice or activity system. This arises from resistance to the new model of the practice or activity, which is distinguished from the old one by its object and associated goals. These contradictions may also be associated with *unlearning* old ways of doing things and implementing new tools or principles. For example, agricultural scientists who are used to telling farmers what to do and not valuing what they say may find it difficult to implement the principle of respecting plural ways of knowing. Overcoming a tertiary contradiction results in the vertical movement of an activity system or practice. When the tertiary contradictions are resolved, quaternary contradictions emerge.
4. *Quaternary contradiction.* Quaternary contradictions arise when the central activity system has improved internally to become historically more advanced but encounters obstacles in relating with those that it interacts with. For example, if we

treat Bouwas Mawara family as an activity system that improved its agricultural practice, we see that resistance came from some of the fellow villagers. His most strategic response was to form an innovative farmers' network which adopted the new way of doing agriculture and whose object went beyond household food security. This ensured the horizontal movement of the innovative practice, which benefited the wider community, in which his family existed. At a larger scale, a quaternary contradiction is overcome, when, for example, the development of a new or improved agricultural practice such as organic farming or conservation agriculture is incorporated in government policies, technology development, curricula, extension work, farmers' practices and national budgets. Resolving quaternary contradictions therefore leads to scaling up and out of innovations.

The progressive emergence and resolution of contradictions of a developing activity system in the course of expansive learning is described below – as logical phases of expansive transformation (Figure 3.2). The developmental work research needs to remember that when he/she arrives at the scene the contradiction of primary concern may be secondary, tertiary, quaternary or primary. The starting point of an intervention is therefore not necessarily always a primary contradiction.

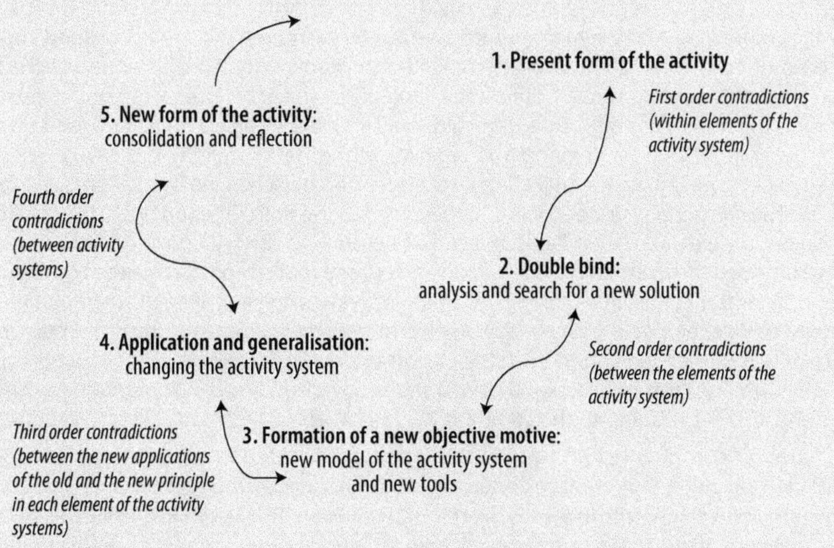

Figure 3.2. Sequence of contradictions during the course of expansive learning and transformations of practices (Engeström, 2014).

Again, it is important to recognise that farmers, scientists and other people encounter and attempt to resolve dilemmas in their respective practices and organisations in their everyday lives. DWR should build on these developments and enhance them by providing means by which the processes can be enriched and advanced more swiftly and deeply. Textbox 3.3 tells the story of a farming community that encountered and resolved an agricultural dilemma to illustrate the point.

Textbox 3.3. The story of a farming community that faced and resolved a dilemma.

An agricultural community in the Mashonaland East Province of Zimbabwe had historically kept cattle for draught power, meat, milk, manure and for traditional functions. One of the traditional practices was that each family had to keep a bull that was named after the great grandfather of the family partly to honour him; and partly to have his spirit protect the family through the bull. This had the practical value of enhancing the calving rate of cows. But this practice started dying down in the 1970s as most people became Christian. Consequently fewer families kept bulls as they were also notorious for straying into people's field, destroying crops. The negative effect of this was that there was a smaller pool of bulls to keep the community's breed healthy. Gradually, the quality of the breed deteriorated. It was smaller and less able to provide draught power, quality meat or high volumes of milk. The community realized that there was need to introduce more bulls and the more progressive farmers acquired bulls, which helped improve the community's breed. This improvement created a problem for the bull owners as the bulls would often follow cows to their different villages. But when the cows were put in kraals for the night, the bulls were left out and would stray into the villagers' fields and feed on their crops. The villagers would demand and receive compensation for the destroyed crops, yet they were benefiting from the bulls. This created tension between the bull keepers and fellow villagers. The community understood that they needed the bulls for good breed but that it also needed good yields from their fields. It faced a dilemma – a catch 22 position. In the end, the community met and examined the situation and agreed that the old law was not good for the current circumstances because it punished bull keepers who were serving the community. They subsequently developed a new by-law which compelled villagers to protect the bulls in the community by either keeping them in their kraals or driving them back to their owners' place. From then on, any crop destruction that was caused by a bull would not be compensated for. In the end, this contributed to the community's improved cattle breed and crop production.

If the farming community had used a DWR approach located in third generation CHAT it would have involved farmer representatives, veterinary department extension workers and the livestock research unit to jointly work out a solution. It is very likely that the solution would have involved some form of participatory animal breeding. Artificial insemination, a technology that is relatively cheaper and accessible in the country, is likely to have been scrutinized as one of the potential solutions thereby widening the pool of available tools.

3.6 Key principles that underpin DWR

In their paper on methodological demands of formative interventions based on DWR and CL workshops, Engeström *et al.* (2014) identify three key principles, namely: *double stimulation; ascending from the concrete to the abstract*; and *transformative agency*. Each of these three principles is briefly discussed below.

3.6.1 Double stimulation

> *Double stimulation is a multi-layered and longitudinal process in which both the initial problem situation (first stimulus) and the mediating conceptual tool (second stimulus) are reformulated and enriched in successive steps. Such a process of double stimulation generates a thirdness, a new concept for the activity under transformation.*
>
> *(Engeström, 2011: p. 1)*

The principle of double stimulation provides a generative mechanism for change (Engeström *et al.*, 2014). Through it people intentionally break out of a conflicting situation and change their circumstances (Sannino, 2011). The starting point of double stimulation is a conflict of motives, which may manifest themselves through dilemmas or contradictions.

The first stimulus is the problem to be tackled, which may initially be expressed as a tension, conflict disturbance, rupture and later be framed as a contradiction. It is a problem that the subject cannot solve alone with the help of previously learned concepts and methods while the second stimulus is a neutral tool that the subject can make into an instrument for organising the problematic situation to develop a solution (Virkkunen & Schaupp, 2008). For example, in her study of Ethics-Oriented Learning in Environmental Education in South African Workplaces Olvitt (2010: p. 84) identifies the following contradiction, which showed conflicting motives between curriculum designers and learners: 'The course's written materials and assessment tasks are experienced by some learners as inaccessible, even alienating'. Seppänen (2004) identifies the contradiction between *'short-term and intensive use of natural resources against their ecological and sustaine use'* in the production of organic vegetables in Finland.

Getting out of the difficult situation using double stimulation involves the formation of new concepts, solutions and skills and associated transformative agency among participants (Engeström, 2007) based on the utilisation of a second stimulus. It is a neutral tool that the subject can make into an instrument for organising the problematic situation towards developing a solution. It is an external artefact that the subject turns into a psychological tool and uses to (re)structure the first stimulus understand it better and control subsequent actions. It sparks off a fundamental problem-solving process (Virkkunen & Ristimäki, 2012). Although initially the CL researcher provides the second stimulus, it is the research participants who should provide one ultimately in order for the intervention to be effective. The activity system and the expansive learning process have been used as the initial second stimulus in many DWR interventions. The activity system is typically used for the framing of human activity and the location of contradictions. The efficacy of the second stimulation lies in its ability to enable participants to break out of the problematic situation. The critical role of the developmental work researcher here is to help research participants articulate their problematic situation; and then to find the conceptual tool or tools that can help them understand the situation better and break out of it.

3.6.2 Ascending from the abstract to the concrete

> *This process of ascending from the abstract, isolated germ cell of a system to a multi-faceted, concrete reality can be easily understood as the process of coming into being of a new technical system.*
>
> *(Virkkunen & Newnham, 2013: p. 45)*

DWR is primarily concerned with theoretical abstraction, which is based on practical transformations, experimentation and the altering of relationships in a system of interdependent elements. This is achieved through the invention, construction or modelling (in the mind) of an idealised design or scheme of action (Davydov, 1990). The process of resolving the inner contradictions involves double stimulation, the development, examination and improvement of model solutions through creative experimentation, and the negotiated application of the successful solution in other situations (Davydov, 2008). As Spielman *et al.* (2009) note, such a creative solution only becomes an innovation when it is successfully put into practice in a concrete environment. Distinguishing concrete and abstract in dialectics, Tolman (1981: p. 42) says:

> *The key to the correct understanding of the dialectical alternative lies in understanding of the word 'concrete'. The concrete understanding of the thing is an understanding of it in its interconnections and movements. To extract a thing from its interconnections is to abstract it.*

The meaning of concrete used in the above sense resonates with AIS's integrated approach, which recognises the interconnectedness in agricultural systems and value chains. The principle of ascending from the abstract to the concrete is embedded in the DWR process since:

- The starting point is contradictory demands that are faced by research participants in their course of work. These occur in the real concrete world of the farmer, the natural resource user, agricultural researcher, extension worker, food processor, marketer, consumer, etc. This obstacles are identified through questioning the current wisdom of an activity or practice;
- The next stage comprises the abstraction of a principle or a 'germ cell' that gives rise to the development of a new or improved activity. The germ cell can exist in the form of a new principle or an exemplary new kind of action that embodies the new principle (Virkkunen & Ristimäki, undated); and
- The third stage involves application, enrichment and expansion of the principle, germ cell or innovation or new solution in concrete and real situations. This application first affects the immediate activity system and later affects other interacting activity systems, transforming broader community or societal practice. This process reaches its peak when the innovation achieves internal completeness and stability, external necessity and universal generalisation. The spread of the conceptual tool or innovation is also called theoretical generalisation and takes place through relational and collective agency (Engeström, 2011: p. 606; Virkkunen & Newnham, 2013: pp. 44-46).

3.6.3 Transformative agency

> *Transformative agency differs from conventional notions of agency in that it stems from encounters with and examination of disturbances, conflicts and contradictions in the collective activity. Transformative agency develops the participants' joint activity by explicating and envisioning new possibilities. Transformative agency goes beyond the individual as it seeks possibilities for collective change efforts ... Transformative agency also goes beyond the situational here-and-now actions as it emerges and evolves over time, often through complex debates and stepwise crystallisations of a vision to be implemented.*
>
> *(Haapasaari* et al., *2014: p. 2)*

While double stimulation is concerned with how problematic situations are resolved in DWR and ascending from the abstract to the concrete explains how activity systems and practices are improved and expanded through DWR, the principle of transformative agency describes what DWR participants do to cause change –

their agentive actions. There are at least five types of agentive actions that are likely to emerge in the course of a DWR intervention (Engeström, 2011), and these are:

- Resisting and intervening through criticising and questioning current wisdom or practice;
- Explicating new possibilities or potentials by, among other things, drawing from past positive experiences and acknowledging their potential as a source of new possibilities;
- Envisioning new patterns or models of the activity, which is concerned with the development, presentation and examination of model solutions;
- Committing to concrete action aimed at changing the practice, where the actor expresses his/her intention to act in a specific way; and
- Taking consequential action to change the way things are in real life situations, often beyond and outside the CL workshops.

The evolution of the above agentive actions is accompanied by learning among those involved. The five types of agentive actions also tend to show a shift from individual initiatives to more collective forms of transformative agency (Haapasaari *et al.*, 2014). This is consistent with the idea of developing individual, relational and collective agency in CHAT and DWR, as discussed in Chapter 2. The developmental work researcher is encouraged to distinguish and pay attention to all forms of agentive action in facilitating change processes as they determine the success of collaborative learning, innovation and action. The most effective way of doing this is to video-record the CL workshop discussions and review them later.

3.7 Change laboratory workshop: the spine of DWR

The CL workshop is an interventionist research method that was developed by the Centre for Activity Theory and Developmental Work Research at the University of Helsinki in the mid-1990s in the context of DWR and CHAT. Moldashl & Brödner (2002, cited in Pihlaja, 2005: p. 190) note that there are three models of intervention: the expertocratic which assumes that scientific knowledge is superior to practitioner's knowledge; the proceduralistic model which assumes that the knowledge is already available in the organisation and only needs to be mobilised; and the reflexive model which transcends the two and stimulates re-negotiation and self-reflection among both organisational members and external experts. DWR and the CL workshop are based on the reflexive model, which recognizes that some of the needed knowledge resides among the research participants in the area concerned but that some of it may be outside the group and needs to be sought.

3.7.1 Intended outcomes

The CL workshop method is underpinned by the principles of double stimulation, descending from the abstract to the concrete and transformative agency (Engeström *et al.*, 2014). The spirit of a CL workshop is to enhance the building of shared views of the shared object in and between activity systems in order to develop new practices, tools and models (Ala-Laurinaho & Koli, 2007) and associated transformative agency. R. Engeström[1] (2003: p. 2) notes that the 'change laboratory method focuses on historically emerging tensions and contradictions in the activity system' and the goal of the method is 'to expand the understanding of practitioner's activity through experimenting and reflecting' as well as 'promote people's possibilities to utilise their multiple understandings and identities'. The four main outcomes of formative CL workshops are as follows:

1. knowledge and understanding of developmental challenges and possibilities of the activity in question;
2. collective and individual transformative agency;
3. formation of new concepts and invention of new tools and organisational forms; and
4. change and development in the activity and in individuals (Virkkunen & Newnham, 2013: p. 166).

3.7.2 Preparing for CL workshops

There are five main planning stages that are conducted in preparation for CL workshops:

1. *Negotiating the object of the intervention.* As part of negotiating access into the research site and bounding the case study for intervention, the developmental work researcher engages decision-makers in the identified workplaces to agree on the activity systems and/or practice to be examined and improved upon. The object of each activity system that is to participate in the study should also be defined and agreed upon at this stage. For example, Mukute (2013) shows that a developmental work researcher was engaged to work with 32 newly established organic farmer associations whose object was to build their respective human and natural resource capacities to produce organically and access appropriate markets. The 32 organic farmer associations comprised the activity systems; organic agriculture was the practice; and the farmers in the association and the land their agricultural land were the objects that had to be worked on in order to produce organic food that was safe and nutritious for household consumption

[1] This paper was written by R. Engeström, unlike the rest of the papers.

and for consumers who would buy it on the market. Producing organically also entailed producing food and building the ecological capital such as land, water and biodiversity at the same time. The farmers had historically used traditional and conventional agricultural practices whose objects did not explicitly include safe nutritious food and production of ecological services. Although the contract did not explicitly define CL as the intervention method, it provided for the use of Participatory Action Research which would cause learning and enable the production and marketing of quality organic produce (*ibid.*). In addition, the contract also agreed on who was to take part in the CL workshops: members of each association and the local government agricultural extension workers who worked with the associations.

2. *Generating mirror data.* Once the object of the intervention is agreed on, the developmental work researcher generates data with research participants through methods such as interviews, group discussions and historical timelines. These methods are complemented by secondary data from documents and literature on the practice and activity system that is under review. The developmental work researcher processes the data and presents it as mirror data during the CL workshop.
3. *Defining the unit of analysis.* The mirror data generated through the above process provides the developmental work researcher with a sense of the scale and scope of the problematic situations that need to be dealt with. For example, if there are tensions being experienced between activity systems, it makes sense to work with a unit of analysis that enables their simultaneous engagement and this has a bearing on the planning of the CL workshop in terms of who needs to be invited. But more importantly, it has a bearing on the nature of the solutions that need to be generated in order for them to be effective and lasting. A unit of analysis in DWR is the singular entity which exhibits the essential relations of the whole process – essential relations that will stimulate further development and further interactions with other processes (Blunden, undated). He (ibid.) urges developmental work researchers to take time to identify the right unit, which may change with a deeper understanding of the situation. For example, in the Isidore Organic case study in South Africa that resulted in the formation of a Green Growers Association, the basic unit of analysis had to include the organic farmers, trainers (facilitators) and entrepreneurs (Mukute, 2010a: p. 224). This was based on the mirror data that is summarised in Table 3.2.
4. *Developing a schedule of CL sessions.* The developmental work researcher and the client organisations jointly develop a schedule of CL sessions and what each session is intended to cover. The details of the agenda may be changed during the course of the CL sessions if necessary. Table 3.3 shows a sample for a planned CL workshop.

Table 3.2. Summary of contradictions in Isidore case study.

Contradiction	Organic farmers' activity system	Organic facilitators' activity system	Organic entrepreneurs' activity system
Primary	• Between time needed to make money and time needed to build the productive capacity and safe production of food • Between the ecological, economic and social objects of the practice		
Secondary		• Between the pedagogical strategies employed and farmer learning and practice needs	• Between seasonality of local vegetable production and daily demand for them
Quaternary	• Between messages agricultural messages brought by conventional agriculturalists and those brought by sustainable agriculturalists • Between external organic farming rules and the local socio-ecological conditions (rules) of farmers and farming	• Between time and resources allocated for supporting farmer learning and time and resources needed for it	

5. *Detailed planning of CL sessions.* The detailed planning of the second and subsequent CL sessions is built upon the achievements and analysis of the previous session. This means that the schedule of sessions may be adjusted. The developmental work researcher should closely study and analyse the video-recorded material or notes taken on the previous session and use this to produce mirror data and other important outcomes. For example, during the first five CL sessions in the Isidore Organic Agriculture case study referred to above, mirror data generated during the workshop revealed that the unit of analysis needed to include organic inspectors who would assist in the certification of the farmers' organic produce. Important parts of the video-recorded material need to be transcribed for later use – such as for analysing the emergence of agency. This will also guide the developmental work researcher to recommend the conceptual tools to work with in the next session. The importance of well-documented CL workshop sessions is underlined by Virkkunen and Newnham (2013: p. 80):

Table 3.3. Sample CL sessions plan.

Session	Focus	Main processes
1	Orientation to the workshop and review of current practice	• The researcher explains the background to the workshop, the planned process and the intended outcomes • Participants review their practice and reveal obstacles, disturbances and dilemmas
2	Identification of (shared) object and contradictions	• Research participants define their shared object • The researcher presents 'mirror data' that participants engaged with and enrich • Groups *prioritise* the difficult and complex issues (contradictions), *negotiated* and agreed on a few to work on
3	Analysing contradictions	• Research participants analyse contradiction in terms of their *evolution* (past and present) and their *causes and effects* • Research participants agree on the contradictions to work on
4	Developing model solutions	• Research participants *jointly constructed* model solutions in groups • Research participants share model solutions in the plenary for *critiquing* and improvement • Produce model tools that have potential to overcome the selected contradictions and improve the practice
5	Way forward	• Develop structures and a plan for implementing and improving the model solution(s)
Implementation		• The developmental work researcher compiles and submits a report on the proceedings to help research participants to move forward • Research participants implement the model solution and make some adjustments to it based on realities on the ground • Research participants identify and record obstacles arising from the implementation of the model solution
6	Feedback and review of solution(s) – six months after 1st session	• Research participants report on the progress that they made in implementing the model solution • Identify emerging contradictions that need to be resolved
7	Modelling of solutions to new and emerging contradictions	• Develop and examine model solutions to identified tertiary contradictions
8	Way forward	• Develop a plan for implementing and improving the model solution(s) to address tertiary contradictions
Implementation		• Research participants implement new and revised solutions and make necessary adjustments • Research participants identify and record obstacles arising from the implementation of the new and revised solutions • Client identifies, mobilises and briefs representatives from affected interacting activity systems so that they become research participants

>>>

Table 3.3. Continued.

Session	Focus	Main processes
9	Feedback & review of solution(s) – 18 months after 1st session	• Research participants report on the progress that they made in implementing the model solution • Expand the (shared) object that incorporates interacting activity systems • Develop and examine solutions to emerging quaternary contradictions
10	Way forward	• Develop an inclusive structure and a plan for implementing and improving the model solution(s) to address quaternary contradictions

> *The central material in the planning of a session is the previously collected data about the activity and the researcher-interventionists' preliminary hypothesis concerning the central problems and inner contradictions in the activity system. It is good practice to videotape each session for immediate and later analysis. The researcher-interventionists can analyse how the expansive learning process progressed in the session by analysing the video recording. They can analyse what important themes arose and got the participants involved, as well as what manifestations of inner contradictions in the activity system appeared in the participants' talk during the session ... On the basis of the analysis the researcher-interventionists decide what themes need further elaboration, what kind of additional data would perhaps be needed, and what kind of analytic concepts might help practitioners [and specialists] to progress in their expansive learning process in the Change Laboratory.*

In cases where video-recording is not feasible, it is still important to keep a detailed record of the proceedings for subsequent analysis. It is therefore important to have someone responsible for taking minutes during the proceedings – preferably someone who understands the practice or activity system(s) under review.

3.7.3 The CL workshop setting

The typical CL workshop layout is depicted in Figure 3.3. The diagram shows that a CL workshop involving multi-stakeholders should ideally be attended by practitioners and subject or thematic specialists and facilitated by a developmental work researcher who should be accompanied by at least one other developmental work researcher. The number of practitioners and subject specialists will vary according to the situation but 12-30 people seem to be ideal. In addition, someone should video-record the workshop proceedings while another takes notes. The

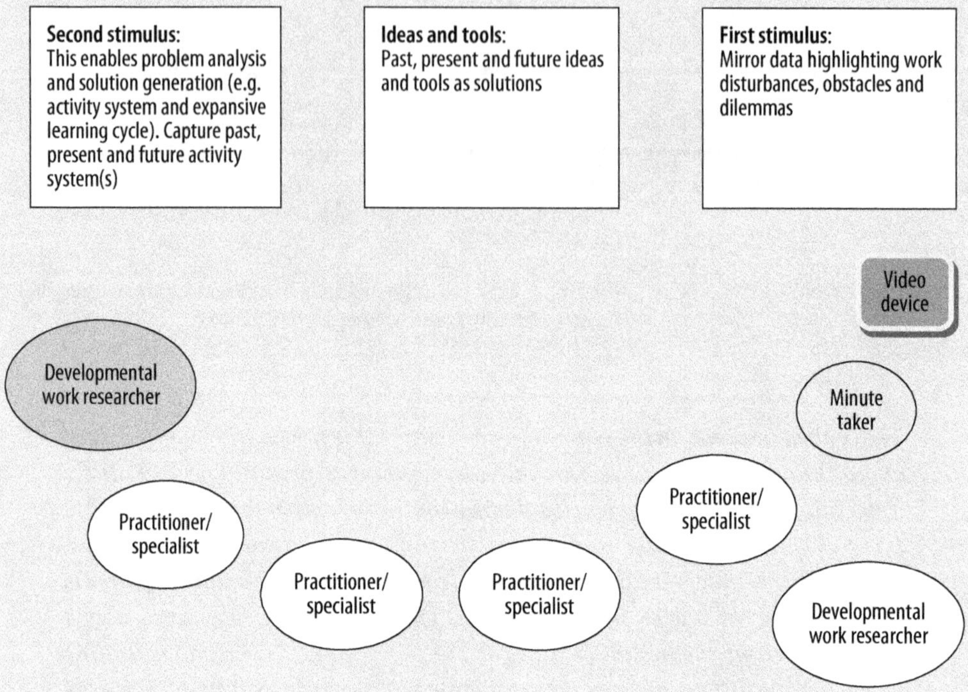

Figure 3.3. CL workshop layout and tools (Engeström *et al.*, 1996).

workshop uses three surfaces to visualise workshop proceedings. The top right hand corner surface is used for capturing 'mirror data' being the work disturbances and problematic situations, and relevant novel innovative solutions that are already in circulation. The information is collected from interviews, stories, documents and conversations during the workshop. The second stimulus, which comprises theoretical and conceptual tools that are used for analysing mirror data and looking for solution, is depicted on the surface at the extreme left. The conceptual and theoretical tools that are typically used during the first few sessions of the workshop are the activity system model to help reveal inner contradictions and expansive learning cycle to analyse the current and develop a possible evolutionary path of the practice under review. Research participants may replace the tools with others during subsequent sessions. The third space that is in the middle is used for capturing new ideas and potential solutions to the practice or work disturbances (Ahonen *et al.*, 2000).

3.7.4 Steps in implementing a CL workshop

CL workshops are organised into sessions, which follow the sequence of the stages of the expansive learning process and each sessions tends to last two hours. The sessions that question the practice, surface and analyse contradictions, model and examine solutions should follow each other within a few days in order to build on the momentum of conversations and accumulation of ideas (Virkkunen & Newnham, 2013). Sessions to review the implementation of the model solutions and look for tertiary and quaternary contradictions should be held several months after implementation to allow for substantial experience on the ground, between CL sessions. The total number of sessions that are carried out before implementing the model solutions ideally range from five to twelve (*ibid.*). Their number varies and depends on several factors such as time and resources available, complexity of the contradictions to be resolved, range of actors and activity systems involved, distances to be travelled and negotiations between the developmental work researchers and research participants. For example, Engeström & Sannino (2011) conducted 8 sessions in the University of Helsinki Library case study; Mukute (2009, 2010a) conducted five sessions in each of the three sustainable agriculture case study sites; and Engeström *et al.* (2007) also conducted five sessions in a medical activity system. If sessions are fewer then more work has to be done between sessions and the sessions could be made longer.

Below is a description of the steps to take in designing and implementing a CL workshop (Textbox 3.2). The steps are taken in the broader context of the expansive learning process described earlier in this chapter.

1. *Questioning.* Generate data about the practice and the concerned groups of people through individual interviews, focus group discussions, observations and document analysis. The data should highlight the difficulties being faced and emerging innovations. These difficulties should be organised and presented as 'mirror data' to representatives from the community or communities involved in the practice under review. This stage is primarily concerned with identifying the obvious and visible problems. Table 3.4 gives an example of questions that could be posed to extension workers and farmers respectively in relation of farming before the CL workshop is conducted in order to generate mirror data. While the above questions can be helpful and interview data can help in generating mirror data, it is always important to remember that interviews should build a view of the activity's problematic situation. The video-recording of a narrative of the problematic situation works better as mirror data as it captures the disturbances, hesitations and conflicts that are not yet interpreted. Before seeking information on the problematic aspects of an activity, it helps to ask

Textbox 3.2. How a developmental work researcher could utilise CL workshops.

CL workshop sessions are organised around the steps in the expansive learning process. The developmental research worker initially generates data through ethnographic research. The next step is to conduct historical and empirical analysis. This process should reveal disturbances and innovations in the practice. He or she then presents the findings as 'mirror data' to the people concerned (subjects of the activity systems). The presentation is followed by negotiation of the meanings and identification of the principal tensions. Using the double stimulation principle, the participants develop a joint understanding of the problematic situations and the conflicting motives that they may have. The next stage is the development of model solutions to address the problematic situations. This is achieved through collective experiments. Participants then examine the model solution for improvement. The improved solution gets implemented outside the CL workshops.

During implementation, which takes place outside CL workshops, participants draw on the cognition of people who may not have been present at the workshop. For example, a farmer group may find it necessary to include a soil analyst.

It is advisable for the developmental work researcher to work with a research assistant who is conversant with the DWR methodology. He/she assists with documentation and review of the process. One of the most effective ways of documenting CL workshops is video-recording. The video-recorded and or audio tape recorder material is later analysed for evidence of tensions, learning and agency. If the necessary facilities for video-recording and analysis are unavailable, copious note-taking becomes imperative. The notes are used for subsequent analysis.

practitioners about how they see the purpose and meaning their activity and what satisfies them in it. This helps reveal the variety of meanings and personal senses of the same activity among the practitioners and also the historical layers of understanding it.

2. *Jointly identifying systemic causes of problems in the practice and their origins.* At this stage of the expansive learning process, the participants look for the historical origins of current problematic situations through *historical* analysis. They may collectively construct historical timelines of the practice under review during the workshop. The workshop participants include practitioners from the concerned activity systems and specialists in the thematic areas being covered. A developmental work researcher with the assistance of one or more fellow researchers normally facilitate and video-record the workshop proceedings. The video-recorded materials are subsequently reviewed for evidence of contradictions, learning and agency development.

Table 3.4. Sample questions for extension workers and farmers ahead of a CL workshop.

Key questions for extension workers	Key questions for farmers
• Please explain your work and what you like about it. • Why and how did you learn agriculture? • What challenges did you encounter in learning it? • How do you facilitate the learning of agriculture among farmers? • What challenges do you encounter in facilitating the learning of agriculture? • What constrains or enables the learning and practising of agriculture? • Who are the main stakeholders of agriculture and what roles do they play? • What are the strengths and weaknesses of, opportunities and threats to the agricultural practice that you promote? • How are ecological, social and economic outcomes met in the agricultural practice that you promote? • What new ideas and materials tools are likely to help you in promoting the agricultural practice?	• Please tell me the story of your work as a farmer. • Why did you choose to practise agriculture? • How do you learn agriculture and improve on it? • What gaps do you see in the way agriculture is taught or learnt? • What are the strengths and weaknesses of the agriculture that you practise? • Who are the stakeholders in agriculture and what roles do they play? • What policies and laws encourage or discourage your agricultural practice? • How are ecological, social and economic outcomes met in your practice? • What challenges are you facing in learning, practising or improving your agricultural practice? • What new ideas, seed, breed, equipment and discoveries have potential to improve your agricultural practice?

Participants note the changes in the practice over time and identify the inner contradictions that have been addressed and those that have remained unresolved. They also identify if, when and how the object of the practice changed over time. A historical and systemic view of the activity system(s) involved in the practice is essential at this stage of the process. The following questions could be used to guide the analysis:
- What may have caused the contradiction?
- How extensive is the contradiction?
- In which places and when does the contradiction tend to flourish?
- What are the trends and tendencies of the contradiction?
- What are the consequences of the contradiction?
- What solutions have been developed to address it?

Mirror data, which draws ethnographic research, is used to augment the collective memory of those present. Table 3.5 shows the historical timeline that was developed in the Schools and Colleges Permaculture Programme (SCOPE) with DWR research participants.

Table 3.5. St Margaret School Community Permaculture historical timeline.

Date	Action taken	Object	Problems and solutions
1994	• The Schools and Colleges Permaculture Programme selected St Margaret Primary School as a pilot school for the introduction of permaculture – a sustainable agricultural practice	• Land conservation (stop erosion of school grounds)	• Lack of design skills • Training and involve parents, teachers and pupils in the re-design and improvement of the school grounds
1995-1996	• School and community representatives design the school into permaculture zones • Raising of crops, fruit trees and flowers	• Land conservation, food, beauty and curricula implementation	• Inadequate knowledge and skills on permaculture • Train and engage pupils, teachers and parents in water harvesting and crop production in the school • Water limits production
1997-1999	• Installed a water tank with external assistance	• Land conservation, more food (in dry season), beauty and curricula implementation	• School produced a lot of food from the garden and sold some of it • The school had many orphaned and vulnerable children • Water still the main limiting factor for further expansion
2000-2001	• Introduced a nutrition garden to provide free food for the orphaned and vulnerable children • Used income sale to pay fees for the children • Acquired and installed a water pump	• Land conservation, food and nutrition security for vulnerable children beauty, curricula implementation and income generation	• Too many schools conduct look and learn visits but there is no structured way to support them • Develop a network of schools in the district that learn from each other
2004 onwards	• Cluster system introduced by SCOPE • Teachers from pilot schools such as St Margaret trained to train others	• Land conservation, food and nutrition security for vulnerable children beauty, curricula implementation, income generation and spread of the practice to neighbouring schools	• Local teachers do not have the time, mandate and capacity to train other schools • Train the trainer courses designed and implemented for outstanding teachers • Permaculture resource book for schools produced • Water supply disrupted because of electricity supply leading to decline in production and productivity
2008	• A Change Laboratory workshop was conducted, involving teachers, parents pupils, a local NGO and government agricultural extension officer		

Empirical analysis is used to complement historical analysis so as to reveal contradictions. The workshop can use force field analysis or problem tree analysis to reveal the double bind situations and structural causes of current problems. The activity system framework can also be used to identify the contradictions being faced in a practice or by an activity system. But the simple question that can trigger the revelation of such contradictions could be: *What do you consider to be the critical incidents of failure and success in your activity?*

The historical and empirical analysis enables a deeper and collective understanding of the contradictions or difficulties at hand, and prepares participants to look into the right places for solutions. It may also enable them to identify the right kind of people to invite for solution generation, which is the next stage.

3. *Developing a shared object.* Where the intervention involves more than one activity system, it is necessary for the workshop process to include the collective search and definition of the shared object. For example, in the New Zealand Case Study presented in Chapter 1, the shared object of the New Zealand government and the New Zealand apple industry was a *'A profitable, sustainable legal system of production'* Hill *et al.* (2007: p. 364). Once the shared object is identified, it is necessary to reformulate the identified contradictions in relation to it. In addition, and more importantly, this process should also redefine most important contradictions to which solutions must be sought. Selection of contradictions to work on is normally done through negotiated ranking.
4. *Modelling solutions jointly.* After selecting contradictions to work on, research participants begin to model solutions by drawing on their distributed knowledge, skills, experience and imagination. Model solutions are forms of tools which could be material, conceptual or other people. In modelling the solutions, it is often better for them to go into small groups that allow everyone to have a chance to make a contribution.
5. *Examining the model solution.* At this stage, the model solutions that are developed in different groups during the CL workshops are presented in the plenary. The rest of the participants comment on each model solution in terms of its potential as well as limitations in addressing the contradictions and problematic situation. The potential negative consequences are also examined. This is followed by an improvement of the model solution. The idea behind this process is to pre-test and strengthen the solution before it is tried in the field, using the participants' knowledge and imagination. The following two questions are useful for guiding the discussion:
   - What are the strengths and limitations of the proposed solution?
   - What can be done to improve the adequacy of the proposed solution?

6. *Implementing the solution.* Outside, between and beyond CL workshop, research participants, who have also become innovators, implement the model solution in a real life situation. They also make changes and improvements to the model solution as new insights emerge. The period of implementing a model solution depends on the nature of the solution and the complexity of the difficulty being addressed. It could take place over a few months or a couple of years. It is important that the participants/innovators document the process of improving the solution because it often carries lessons and insights that are useful beyond the immediate community.
7. *Reviewing the model solution.* After implementing and adjusting the model solution for some time, the research participants/innovators and developmental work researcher meet in a CL workshop to jointly review the model solution for possible adoption. During the workshop, the intended and unintended consequences of employing the solutions are made visible. The following questions may be used to guide discussions:
   - How did you go about implementing the model solution?
   - What challenges did you face in implementing the model solutions?
   - What adjustments to the model solutions did you make and why?
   - What intended and unintended effects did the implementation of the solution have?
   - What problems or contradictions have the implementation of the solution causes in relation to neighbouring or stakeholder activity systems?
   - What lessons and insights have you generated through implementing (and adjusting) the solution?

This may lead to the adoption of the tool in its form, or in an adjusted form. It could also result in the rejection of the tool depending on the assessment. If the tool is accepted, it means that a vertical movement of the practice has been achieved and the focus shifts to the next stage – consolidation – beyond the activity systems of the actors who took part in the intervention.

8. *Consolidation.* This stage is concerned with the horizontal movement of the innovation (or the principle behind the innovation) into the other activity systems that have a bearing on the practice under review. It is achieved through identifying and addressing quaternary contradictions which ensures that the innovation achieves societal value because if the contradictions arising from relations with other activity systems are not addressed the success of the innovation will be temporary. For example, if the improvements in the apple industry in New Zealand do not spread to other sectors of the economy, they would not last. Once consolidation is done and the people concerned end up

with a more advanced activity system, they go back to the drawing board to look for potential areas of improving their practice or organisation. This allows them to continuously develop their practice.

3.8 The DWR process: an illustrative example

Scores of interventionist studies have been conducted using DWR in the agriculture and NRM sectors. Some of these examples are listed in Chapter 1. This sub-section shows how DWR was utilised in a particular study that was conducted in Lesotho in 2008 and 2009 (Mukute, 2010a; Mukute & Lotz-Sisitka, 2012). The study focused on exploring and expanding agriculture in the context of sustainable development in Lesotho where some of the challenges being faced in conventional agriculture demand new ways of doing agriculture.

3.8.1 Background

Historically, Lesotho's agricultural practices used to produce enough to feed the nation and generate surplus, which was exported to South Africa. Initially, this was based on traditional agriculture and later on conventional agriculture, which introduced new and improved crop varieties but used deep ploughing methods. But the soils are prone to erosion and this is worsened by steep gradients of the mountainous country. As a result, soil erosion became commonplace, gullies were formed and agricultural productivity fell. Household seed and food security declined, thus impoverishing the rural communities who are dependent of agriculture. Men left to neighbouring South Africa to work in the mines, further undermining agriculture as they were an important source of labour. In looking for ways to transform conventional agriculture to meet local social ecological conditions, a sustainable agricultural practice was developed but it is facing a lot of resistance (Machobane & Berold, 2003; Mosenene, 2000; Robertson, 1994). Still, the need to improve agriculture so that it meets the ecological, economic and social interests of its stakeholders remains important for the agrarian country and its communities. Meanwhile the government budgets, research and extension system largely promote conventional agriculture.

3.8.2 Questioning the practice

The researcher was motivated by his knowledge and understanding of the potential and challenges of agriculture in Lesotho, having worked with several agricultural NGOs and farmer groups in the country. He initially generated data through literature review and document analysis, covering the historical and contemporary periods. He also generated primary data from farmers, agricultural extension workers and an agricultural researcher. The products of questioning the practice were issues and disturbances that had affected the development of agriculture in the country. At the same time, the study showed how the purpose and means of agriculture had changed over time. This is called object-historical analysis. In particular it revealed how one form of agricultural practice called MFS had emerged and was struggling to succeed the dominant form of conventional agriculture. This partly addressed what theory-historical analysis. But the interviews were particularly important in showing up the current obstacles, issues and challenges that farmers and extension workers were facing in agriculture. These included land degradation, household food insecurity, seed insecurity, climate change and inadequate support from government. This comprised the first layer of actual empirical analysis.

3.8.3 Analysis of the problematic situations

A few months after the ethnographic research the developmental work researcher and an assistant researcher met with the agriculture stakeholders in a CL workshop. The CL workshop was attended by sustainable agriculture farmers (who practice the Machobane Farming System – MFS) and agricultural extension workers from two districts in which the research was taking place and an NGO interested in sustainable agriculture. The researcher presented the data that he had identified from the questioning stage, highlighting the disturbances that were being faced in the workplaces of farmers and extension workers. The participants gave their feedback on the mirror data, which was adjusted. Force field analysis and historical analysis were the tools used to generate additional data. For example, the farmers' historical analysis, which was done in the local language and presented on flipchart, is partly shown in Figure 3.4.

The assistant researcher video-recorded the proceedings for subsequent analysis. The developmental work researcher then presented the second generation activity system and the expansive learning process as the tools that were to be used to further analyse the data and the help the farmers, extension workers and development facilitators to jointly define ZPD for agricultural practice in the two districts. Three interacting activity systems were identified, which located the study in a third generation CHAT.

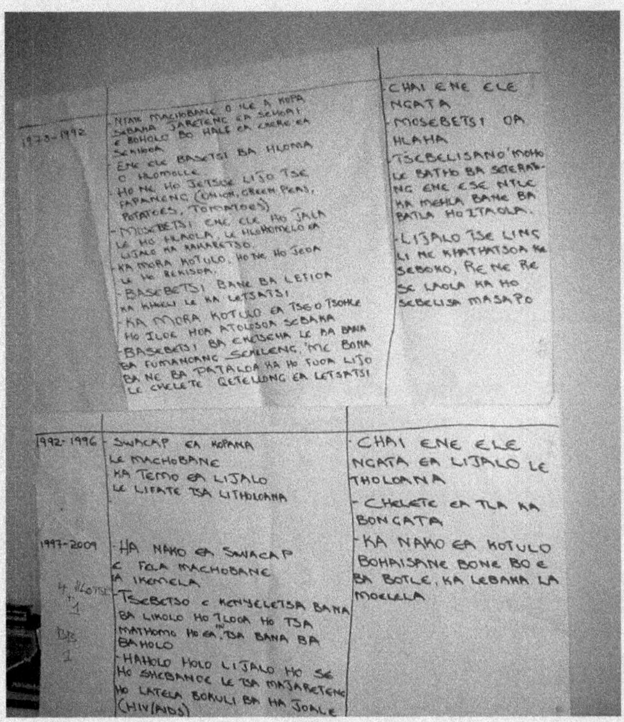

Figure 3.4. Farmers' summarised critical moments and implications of their activity's historical development.

The agricultural disturbances were further analysed for causal explanations. This was aided by the use of a problem tree analysis tool that participants were generally familiar with. The intention was for the subsequent intervention to focus at this deeper level of the activity – not just the operations. The causal mechanisms, which were later reframed into contradictions by the researcher after the CL workshop, included agro-ecological conditions; government, donor and corporate sector policies; lack of sustainable agriculture capacity; land tenure; and uneven power relations between government, donors, research organisations and farmers.

The CL participants prioritised the lack of sustainable agriculture capacity and analysed it further before moving to the next stage of the process. They identified the following issues:

- there was no joint planning between government, farmers and agricultural NGOs;
- lack of national and local government budget allocation for the specific kind of sustainable agriculture that was adapted to the local conditions (Machobane Farming System); and
- exclusion of the sustainable agriculture practice in formal agricultural curricula in schools, colleges and university, which undermined its growth and development despite the potential that it had demonstrated.

3.8.4 Modelling a solution

The farmers, extension workers and sustainable agriculture participants who attended the CL workshop jointly defined the shared object of their activity systems based on what would be possible to achieve if the current obstacles were overcome, and this was: *using land more effectively and sustainably to produce more food and ecological services.*

The joint definition of a shared object is similar to the process of collaborative agenda setting. The formulation of a shared object was important as a way of expanding the local agricultural activity while at the same time capturing the motive of their collaborative learning, innovation and agency. The initial model solution was joint planning meetings between government, farmers and NGOs involved in sustainable agriculture. But this was refined in the next session of the CL workshop.

3.8.5 Examining and refining the model solution

The refined solution had two parts: a multistakeholder learning and innovation platform in one district (Mafeteng) and a demonstration plot where the sustainable agriculture practice was to be implemented. The platform comprised government agricultural extension workers, local government (district authority), local farmers, and two NGOs that promote sustainable agriculture in the form of MFS in the country. The demonstration plot was to be a site where farmers, extension workers and the two organisations were to collaboratively learn, practice, evaluate and improve the practice in line with their shared object.

The modelling of the solution marked the start of the journey of ascending from the abstract to the concrete. Ascension into the concrete situation began after the CL workshop, when research participants began implementing their solution in real-life situations.

3.8.6 Implementing the solution

Within a period of about six months, after the modelling of the solution, the developmental work researcher returned to the site and noted that the following major developments had taken place:

- the Mafeteng district council agreed to participate in implementing the solution. It allocated land for the MFS demonstration plot and fenced it off to protect the cropland from stray animals; and
- the first crop was planted through the joint effort of government extension workers and farmers and MFS promoters.

3.8.7 Reviewing the implementation of the model solution

The reviewing of the implemented solution took place about 6 months after the modelling of the CL workshop. This was before the first crop on the MFS demonstration plot was ready for harvesting. The feedback workshop revealed that, even though the DWR process was incomplete, the following changes had taken place:

- *Learning.* Research participants reported that they had learnt from each other through the platform and the demonstration plot. The developmental work researcher used his insights from this particular case study and two others to propose tools that could be used by farmers to support continuous learning and innovation. Textbox 3.4 describes one of the tools that was developed.
- *Innovation.* The multistakeholder platform provided a space for continued dialogue between farmers, government extension workers, local government and NGOs who had a shared object. This innovation was *socio-institutional* and involved boundary crossing by members of the interacting activity systems. The joint establishment and management of a demonstration plot provided a potential space and process for the joint development of technical innovations towards the shared object. A successful demonstration plot had the potential to influence the larger national government activity system and allow for the subsequent consolidation of the practice.
- *Agency.* Farmers in the study developed a critical awareness of the causal powers of national and local government. The DWR process also created an opportunity for MFS promoters and farmers on one hand; and government agricultural extension workers to discuss together matters of common interest. They got to know each other and listened to each other in a non-threatening environment. This way, relational agency was developed. The historically opposed groups of actors also influenced each other, especially in the development and examination

Textbox 3.4. Farmer learning and development forum.

What became increasingly clear from the study was the fact that farmers and farmer workplaces do not provide structured and regular moments for learning and development of the practice ... In addition, unlike in structured work environments, farmers in communal settings work independently and separately and their workplace learning needs to be conceptualised in this context ... It is against this background I propose Farmer Learning and Development Forums as a social tool ... A forum would serve the following functions, among others:

- Farmers share individual experiences, innovations and seed for individual and collective horizontal learning;
- Receive instruction on topics that they would have identified prior to the meeting from fellow farmers and others to facilitate vertical learning;
- Pose questions and share challenges that they are facing in learning and implementing agriculture to facilitate systemic engagement with issues;
- Collectively analyse the performance of their practice with a view to identifying weak linkages (covering social, economic and ecological) as well as other constraints to exercising critical thinking;
- Jointly and regularly reflect on challenges they encounter and develop ways to address them and examine the potential impact of their proposed solutions through innovative thinking. The solutions would include finding other people to help them develop solutions as well as drawing on indigenous knowledge, local culture and exogenous knowledge;
- Receive and discuss information about new and relevant technologies from elsewhere to draw out what is relevant and useful;
- Receive and discuss relevant local, national and international information that has a bearing on their work to be responsive to contemporary and emerging developments as well as link up with other relevant actors;
- Develop an agriculture research agenda for the area to ensure the relevance of agricultural research; and
- Develop general strategies to productively engage with key stakeholders such as government, policy-makers, agricultural institutions, NGOs, the private sector and consumers to make the best use of available social capital.

Source: Mukute (2010a, pp. 360-361).

of the model solutions. Their common position and shared object enabled them to mobilise their individual agencies into collective agency.

3.8.8 Reflections on the study

A major weakness of the interventionist study being reported here is that it was too short to allow for the whole expansive learning process to be carried out. The three phases into which the research was divided – ethnographic research; CL workshops up to the solution modelling stage; and feedback sessions were conducted in a period of just over one year. This is too short to allow for a new solution to be fully integrated into a practice and for the practice to have adequately influenced its neighbouring activity systems. Good practice in DWR research suggests that interventions should ideally take a minimum of two years. Continuation of the interventionist study is likely to have produced resistance and associated responses intended to make it better (vertical movement) and spread into other activity systems (horizontal movement). Horizontal movement could have meant spreading into other districts such as Mohale's Hoek; and inclusion of MFS in agricultural curricula.

3.9 Microanalysing CL session transcripts to trace agency development

Video-recorded material can be used to trace and understand whether and how agency development grew and developed in CL workshops. Drawing on the work of Sannino (2008), Engeström (2008), Edwards (2007) and Lister (2004) the following five types of agentive talk have been used in analysing agentive talk in CL workshops:

- *Agentive Talk Stream 1 – Suggesting that something is doable/optimistic talk.* This stream involves suggesting that something is doable, using retrospective experiences or prospective possibilities (Sannino, 2008). It can be associated with explicating new possibilities or potentials by drawing from the past positive experiences (Engeström, 2008).
- *Agentive Talk Stream 2 – Reaching mutual understanding.* When different perspectives, voices and knowledges converge and participants agree on something, the ground for collective action is prepared. The bonding of ideas and bridging of distances between participants prepares their disposition to act. This bonding is linked to Edwards' (2007) notion of relational agency and to Sannino's (2008) reciprocal discursive support, which occurs when individuals express a shift from personal to joint action.
- *Agentive Talk Stream 3 – Envisioning new models.* This happens when the object of the activity is re-conceptualised and expanded and invested with new meaning or when the activity system is re-formulated to something more advanced as Engeström (2008) pointed out as the third form of agency. It is also consistent with the idea of constructing a zone of proximal development (ZPD).
- *Agentive Talk Stream 4 – Commissive talk.* This happens when participants commit themselves to taking specific action towards a jointly defined and shared object. It is a kind of talk that conveys the will to act, a commitment to action (Engeström, 2008; Sannino, 2008).

- *Agentive Talk Stream 5 – Awareness of critical[2] activity systems.* This stream draws on the work of Lister (2004) who argues that an awareness of those with power and how such power may enable or constrain change towards the participants' shared object is a significant dimension of agency. This was a particularly useful stream of agentive talk analysis given the cultural historical and ecological context of southern Africa discussed earlier.

Table 3.6 shows how transcribed data from the Isidore CL workshops was used to gain insights into how the five types of agentive talk evolved along the expansive learning process. It is based on speech nearly 500 speech turns that were recorded (Mukute, 2010a).

Table 3.6: Extracts of data showing the nature of agentive talk in the Isidore CL Workshop.

Type of agentive talk		Evidence of agentive talk in the workshop
Category 1: Suggesting it is doable, optimistic talk		
	Sequence 1.1	40. So what geographical coverage are we looking at?
		41. I think that we should rather look at Natal [KwaZulu-Natal] rather than at national level.
		42. What is the opinion of others?
		43. I think if we can take it down to Durban, be area specific, just that ... because this area is fragmented.
		44. I think we will all have a lot more to contribute to Durban specifically.
	Sequence 1.2	69. Basically we have Earth Mother Organic as an established retail store and it is recognised also as a place of reference where people, the public and the market sector we [are] catering for ... can afford. We can meet and see something solid and it is a forum to actually sell organic produce. And it can be a model for other things.
	Sequence 1.3	125. I think one of the weakest links is nutrition. It basically could help as a means, perhaps as a single most important contributing factor to the growth of organics industry in UK. They target the housewives and young mothers, sensitise them to health issues.

>>>

[2] Critical is used here to refer to those activity systems that have a causal effect on the activity system under discussion. This covers tool, rule, community, subject, division of labour and object producing activity systems.

Table 3.6: Continued.

Type of agentive talk Evidence of agentive talk in the workshop		
Category 1: Suggesting it is doable, optimistic talk (continued)		
	Sequence 1.4	192. There is definitely the need for a forum amongst retailers I think we identified from our side that the problem of linkages comes out because of the lack of a forum, a lack of networking ...
		194. We identified that identifying tangible goals and processes especially a common need between the producers and retailers is definitely a way of sorting out these linkages. But the real issue at this point is a tangible forum.
Category 2: Reaching mutual understanding (note that sequences of talk/units of meaning making are significant here)		
	Sequence 2.1	283. I think what I'm hearing at this stage is as a starting point we want to work with a group of people at various levels along the agricultural production and distribution and marketing chain in terms of organic production ...
		285. Then I think once that is established and functioning we can begin to engage and possibly recruit other actors but in the beginning what would be important that we have fairly clear idea ... we have as a group of the converted in a sense.
	Sequence 2.2	370. It seems to be more like a business strategy, an enterprise development.
		371. Yes, that's what X said.
		372. Yes, I think there is a dimension of business and there is also a dimension that is beyond business.
		373. Social investment, that is what we were talking about.
		374. Yes, yaa.
		375. I would like to say we can buy into it and then co-create the identity.
	Sequence 2.3	420. Where do we meet?
		421. That's to be discussed but I don't think that there is going to be a safe place.
		422. Because we have some venue.
		423. The farm here, this is a productive place to get people together because of the fact we are doing it. This is it. And it has a massive impact on the workshop ... so will continue to use this. Part of my plans, I am moving this workshop [meeting place is normally used for carpentry] out of here on Monday to a factory in Pinetown. I'm freeing this place up so that this place so that it becomes more user friendly for what we are doing not only production but also in a production of this organic network so this would be a primary venue I suppose we call it.
		424. I was going to suggest de-centering because we are going to have stakeholders to work with and you continue meeting at one place, it also creates, bring in dynamics so you may have your primary venue but it would be good to have something ...
		425. Yaa, yaa, yaa [Yes, yes, yes].

>>>

Table 3.6: Continued.

Type of agentive talk		Evidence of agentive talk in the workshop
Category 3: Envisioning new models/ reconceptualising the object		
	Sequence 3.1	73. We need a vision to establish a platform of who are interested in practising organic farming to share information which enhances the farming practice. So this platform needs to be for people who are meeting each other regularly learning from each other to do better in what they do, communicate their practice rightly ...
		85. I am going to call it Noah's Ark ... It is very important that it has a label. It has got to have a mark, recognition mark, Noah's Ark.
	Sequence 3.2	231. Yesterday at the end of our discussion we basically agreed that we should set up an organic forum and we agreed that it would be at district level. We agreed who should be part of the forum and what role they would play not only in terms of individuals it can also be groups or something like that and then the next question is what steps should be taken to set up the forum and by whom and if possible also look at the when, when we are supposed to do what. Then if we still have time we can discuss what problems we are likely to face in setting up the forum.
		232. We must think of it as a constructive body rather than a loose arrangement. It has to be an entity ...
		235. The different people we invited, they all see the different components of the forum. Let's hope that they came back.
		236. That's what we are setting up, it is a vehicle to bring and keep relationships together.
	Sequence 3.3	432. The forum has to be target specific. It has to be structured like what we did with the process of this workshop. We started with 15 different things and created a hell lot of talking and excitement, a lot of interest but it took us four days to get down to one thing that we can work with ... In my opinion it needs to be taken one at a time, like for instance the lay knowledge of farmers, that needs to be a target of the forum so that we approach all the stakeholders from the internal ring and the outer ring, put through our networking possibilities of e-mails, telephones, letters and everything ...
		433. And there will always be a section at the beginning of the meetings where one could do follow up of what happened at the one before, like a short summary of what came out of it.
		434. We can conclude that once you have done that, it's the job of internal group of the forum to then disseminate that information back out to everybody.
Category 4: Commitment to action		
	Sequence 4.1	186. What we are saying is that individuals such as ourselves are going to be responsible for establishing infrastructures.
	Sequence 4.2	218. I would like to establish another venue that will create a farmers organic market. So just to create more consumer awareness market so we will continue to look for a venue hopefully there we can build our foundation and do our meetings there and sort of try and create a central point of meeting and try and establish this thing that we are talking about the Noah's Ark.

>>>

Table 3.6: Continued.

Type of agentive talk		Evidence of agentive talk in the workshop
Category 5: wareness of critical activity systems		
	Sequence 5.1	55. [Part of a statement cited in category 1] So depending on where you go, and if you go to provincial government, then the organics will fail because of the attitude of the regional department ...
		64. I found that organisations tend to defend the position that exists. Therefore, they offer a lot of resistance to new ideas. So that was part of my agenda as well with regards to government representation ...
		93. Government will climb on board when the organic sector makes itself work, and then government will have an interest.
	Sequence 5.2	251. So I think that I agree that maybe having them [funders] is jumping the gun but what I think if one could present modus operandi that of people are results driven ...
		252. So to sum that up as a forum we need to develop prototype that is not funder orientated.
	Sequence 5.3	70. It is a really interesting thing that we actually did a radio programme on Radio South Africa and the listening power and education that can happen in a little programme ... So getting media involved is very important for linking up the customer and re-educating them about organic and conventional agriculture and what is happening.

The following conclusion can be made about the evolution of agency in the CL workshop under review:

- An organic forum comprising Durban farmers, marketers, trainers and local government is doable. This comprises a new model of doing organic farming, production and distribution in the area.
- It has to be based on a shared object related to the practising and promoting of organic farming, beginning with Durban district. The forum is built around an enlarged and more advanced object of organic farming that emphasises cooperation, coordination and communication across related activity systems.
- The forum has to be spearheaded by a structure that we set up, and that meets regularly to discuss and make decisions towards realising the object of the forum. The structure can also be seen as a tool that captures individual, relational and collective agency to drive the activity system forward through vertical movement of the practice and activity systems.
- The founding members commit themselves to taking the necessary actions to establish and operationalise the forum, including engaging with key stakeholder groups and securing their buy-in. These should include government, funders and media. The commitment and awareness of critical activity systems provides an

important basis for consolidation of the coordinated approach and associated technical improvements, which should help in the horizontal movement of the innovation across boundaries.

3.10 Conclusions

This chapter elaborated the DWR methodology and illustrated how it can be applied. In the process, it showed how the principles of: ascending from the abstract to the concrete; double stimulation; and transformation agency can be realised in the DWR journey along the expansive learning process. One case study was used to illustrate how a developmental work researcher worked with the methodology in an agricultural context. In order to get a deeper and broader understanding of the process, it is important to read other studies that have been conducted using DWR. A list of these is given in Chapter 1. The next chapter (Chapter 4) discusses some of the insights generated from working with DWR.

References

Ahonen, H., Engeström, Y., & Virkkunen, J. (2000). Knowledge management – the second generation: Creating competencies within and between work communities in the competence laboratory. In: Y. Malhotra, (Ed.), *Knowledge management and virtual organisations* (pp. 282-305). London, UK: Idea Group Publishing.

Ala-Laurinaho, A., & Koli, A. (2007). Disturbance and analysis as a springboard for understanding service concepts. In: *Conference Proceedings on Researching work and learning, 5th International Conference on Researching Work and Learning (RWL5).* December 2-5, 2007 (pp. 26-32). Cape Town, South Africa: University of Cape Town, Human Sciences Research Council, University of the Western Cape & SAQA.

Blunden, A. (undated). *The germ cell and the unit of analysis.* Available at: http://tinyurl.com/nlnuukb.

Bodrožić, Z. (2008). *Post-industrial intervention: an activity-theoretical expedition tracing the proximal development of forms of conducting interventions.* Helsinki, Finland: University of Helsinki.

Brall, S. (2007). Integrating learning and knowledge management in the university. In: *Conference Proceedings on Researching work and learning, 5th International Conference on Researching Work and Learning (RWL5).* December 2-5, 2007 (pp. 90-96). Cape Town, South Africa: University of Cape Town, Human Sciences Research Council, University of the Western Cape & SAQA.

Bronfenbrenner, U. (1977). Toward an experimental ecology of human development. *American Psychologist, 32,* 513-532.

Danermark, B., Ekstrom, M., Jakobsen, L., & Karlsson, J.C. (2002). *Explaining society: Critical realism in social sciences.* London, UK: Routledge.

Davydov, V.V. (1990). *Types of generalisation in instruction: Logical and psychological problems in the structuring of school curricula.* Reston, VA, USA: National Council of Teachers of Mathematics.

Davydov, V.V. (2008). *Problems of developmental instruction: a theoretical and experimental psychological study.* New York, NY, USA: Nova.

Edwards, A. (2007). Relational agency in professional practice: a CHAT analysis. *Actio: An International Journal of Human Activity Theory, 1,* 1-17.

Engeström, R. (2003). *Change laboratory: A new perspective to teachers' professional development.* Helsinki, Finland: Centre for Activity Theory and Developmental Work Research.

Engeström, Y. (1987). *Learning by expanding: An activity theoretical approach to developmental research.* Helsinki, Finland: Orienta-Konsultit.

Engeström, Y. (1999). Innovative learning in work teams: Analyzing cycles of knowledge creation in practice. In: Y. Engeström, R. Miettinen & R. Punamaki, (Eds.), *Perspectives on activity theory: learning in doing: social, cognitive, and computational perspectives* (pp. 377-404). Cambridge, UK: Cambridge University Press.

Engeström, Y. (2007). Putting Vygotsky to work: The change laboratory as an application of double stimulation. In: H. Daniels, M. Cole & J.V. Wertsch, (Eds.), *The Cambridge Companion to Vygotsky* (pp. 363-382). Cambridge, UK: Cambridge University Press.

Engeström, Y. (2008). *From design experiments to formative interventions.* Paper Presented at the ISCAR Conference. September 9-13, 2008. San Diego, CA, USA: University of California.

Engeström, Y. (2011). From design experiments to formative interventions. *Theory and Psychology, 21,* 598-628.

Engeström, Y. (2014). *Learning by expanding: An activity-theoretical approach to developmental research* (2nd ed.). Cambridge, UK: Cambridge University Press.

Engeström, Y., & Kerosuo, H. (2007). From workplace learning to inter-organisational learning and back: the contribution of activity theory. *Journal of Workplace Learning, 19* (6), 336-342.

Engeström, Y., Kerosuo, H., & Kajamaa, A. (2007). Beyond discontinuity: expansive organisational learning remembered. *Management Learning, 38,* 319-336.

Engeström Y., & Sannino, A. (2011). *Volition and agency in organisations: An activity theoretical perspective.* Paper presented at the Conference of Association of International de Management Stratégique (AIMS). June 9, 2011. Nantes, France: AIMS.

Engeström, Y., Sannino, A., & Virkkunen, J. (2014). On methodological demands of formative interventions. *Mind, Culture and Activity, 21,* 1-11.

Engeström, Y., Virkkunen, J., Helle, M., Pihlaja, J., & Poikela, R. (1996). The change laboratory as a tool for transforming work. *Lifelong Learning in Europe, 2,* 10-17.

Gibbs, R. (1994). *The poetics of mind: figurative thought, language and understanding.* Cambridge, UK: Cambridge University Press.

Haapasaari, A., Engeström, Y., & Kerosuo, H. (2014). The emergence of learners' transformative agency in a change laboratory intervention. *Journal of Education and Work.* Retrieved from http://dx.doi.org/10.1080/13639080.2014.900168.

Hill, R., Capper, P., Wilson, K., Whatman, R., & Wong, K. (2007). Workplace learning in the New Zealand apple industry network: a new co-design method for government 'practice making'. *Journal for Workplace Learning, 19 (6),* 359-376.

Kerosuo, H., & Engeström, Y. (2003). Boundary crossing and learning in creation of new work practice. *Journal of Workplace Learning, 15 (7-8),* 345-351.

Kontinen, T. (2013). A Gramscian perspective on developmental work research: Contradictions, power and the role of researchers reconsidered. *Outlines – Critical Practice Studies, 14(2),* 106-129.

Lakoff, G. (1987). Women, fire, and dangerous things: What categories reveal about the mind. Chicago, IL, USA: University of Chicago Press.

Lindley, D. (2014). *Can expansive (social) learning processes strengthen organisational learning for improved wetland management in a plantation forestry company, and if so how? A case study of Mondi.* Unpublished PhD thesis. Grahamstown, South Africa: Rhodes University.

Lindley, D., & Lotz-Sisitka, H. (2014). *Expansive social learning, morphogenesis and reflexive action in an organisation responding to wetland degradation.* Paper presented at the 30th EGOS (European Group for Organizational Studies) Colloquium. July 2-5, 2014. Rotterdam, the Netherlands: Rotterdam School of Management, Erasmus University.

Lister, R. (2004). *Poverty.* Cambridge, UK: Polity Press.

Machobane, J.J., & Berold, R. (2003). *Drive out hunger: the story of JJ Machobane of Lesotho.* Bellevue, DC, USA: Jacana Books.

Miettinen, R. (2004). The roles of the researcher in developmentally-oriented research. In: T. Kontinen, (Ed.), *Development intervention: actor and activity perspectives* (pp. 105-121). Helsinki, Finland: Helsingfors.

Mosenene, L. (2000). *Lesotho farmer with initiative: Machobane and his farming system.* Paper presented at the conference on Promoting Farmer Innovation (PFI). February 6-11, 2000. Mekelle, Ethiopia: PFI.

Mukute, M. (2006). Making connections: The secret of spreading innovations. *The Harvester Special Edition.*

Mukute, M. (2010). *Exploring and expanding farmer learning in sustainable agriculture workplaces.* Unpublished PhD thesis. Grahamstown, South Africa: Rhodes University.

Mukute, M. (2013). Bridging and enriching top-down and participatory learning: The case of smallholder, organic conservation farmers in Zimbabwe. *Southern African Journal of Environmental Education, 29,* 74-92.

Mukute, M. (2014). *Making farmer innovations matter: the Phiri Award brief.* Harare, Zimbabwe: Phiri Award for Farm and Food Innovators Trust.

Mukute, M., & Lotz-Sisitka, H. (2012). Working with cultural historical activity theory and critical realism to investigate and expand farmer learning in southern Africa. *Mind, Culture and Activity, 19(4),* 342-367.

Olvitt, L. (2010). Ethics-oriented learning in environmental education workplaces: an activity theory approach. *Southern African Journal of Environmental Education 27, 71-90.*

Pihlaja, J. (2005). *Learning in and for production: an activity-theoretical study of the historical development of distributed systems of generalising.* Helsinki, Finland: University of Helsinki, Education Department.

Radjou, N., Prabhu, J., & Ahuja, S. (2012). *Jugaad innovation: think frugal, be flexible, generate breakthrough growth.* San Francisco, CA, USA: Jossey-Bass.

Raelin, J. (2007). Towards an epistemology of practice. *Academy of Management Learning & Education, 6 (4),* 495-519.

Robertson, A.F. (1994). Popular scientist: James Jacob Machobane and Mantsa Tlala. *African Affairs Oxford Journal, 93,* 99-121.

Sannino, A. (2008). From talk to action experiencing interlocution in developmental interventions. *Mind, Culture and Activity, 15,* 234-257.

Sannino, A. (2011). Activity theory as an activist and interventionist theory. *Theory and Psychology, 21,* 571-597.

Seppänen, L. (2004). *Learning challenges in organic vegetable farming: an activity theoretical study of on-farm practices.* Helsinki, Finland: University of Helsinki, Education Department.

Seppänen, L., & Koskimies, H. (2002). Farming across the years: Temporal and spatial dimensions in learning organic farming. In: *Proceedings on IFSA European Symposium on Farming and Rural Systems Research and Extension: Local identities and globalisation.* April 8-11, 2002 (pp. 528-537). Florence, Italy: IFSA.

Spielman, D.J., Ekboir, J. & Davis, K. (2009). Developing the art and science of innovation systems enquiry: Alternative tools and methods and applications to the Sub-Saharan Agriculture. In: P. Sanginga, A. Waters Bayer, S. Kaaria, J. Njuki & C. Wettasinha, (Eds.), *Innovation Africa: Enriching farmers' livelihoods* (pp. 74-85). London, UK: Earthscan.

Tolman, C. (1981). The metaphysics of relations in Klaus Riegel's 'dialectics' of human development. *Human Development, 24,* 33-51.

Tomasello, M. (1999). *The cultural origins of human cognition.* Cambridge, MA, USA: Harvard University Press.

Virkkunen, J. (undated). *The change laboratory: A means to overcome the crisis of collective learning in educational settings?* Unpublished. Helsinki, Finland: Centre for Research on Activity Learning and Development, University of Helsinki.

Virkkunen, J., & Newnham, D.S. (2013). *The change laboratory: A tool for collaborative development of work and education.* Rotterdam, the Netherlands: Sense Publishers.

Virkkunen, J., & Ristimäki, P. (2012). Double stimulation in strategic concept formation: an activity theoretical analysis of business planning in a small technology firm. *Mind, Culture and Activity, 19 (3),* 273-286.

Virkkunen, J., & Schaupp, M. (2008). *From change to development: Expanding the concept of intervention.* Paper Presented at the ISCAR Conference. September 9-13, 2008. San Diego, CA, USA: University of California.

Vygotsky, L.S. (1934). *Thinking and speech. Chapter 1 – The problem and the method of investigation.* Retrieved from: http://tinyurl.com/ob7mlx9.

Chapter 4. Conclusions – Insights from the field

4.1 Introduction

The three preceding chapters have made a case for DWR, explained its ontological and theoretical foundation, discussed and elaborated its design and implementation. This chapter focuses on drawing some insights and lessons from the application of the methodology in a wide range of productive work contexts. Such workplaces include primary, manufacturing and service industries. The service industries include the medical and educational workplaces in Finland, the United Kingdom and Botswana. The primary industries include the agriculture and natural resources management workplaces in Finland, Ethiopia, Lesotho, Malawi, Mozambique, New Zealand, South Africa and Zimbabwe. It has been used by university students as well as by process consultants. In the process of working with the DWR process, some challenges have been encountered; and some lessons have been learnt.

The importance of methodologies that enable practices to develop from dissonance and plural ways of knowing has been highlighted by some agricultural learning and innovation initiatives. For example, the Joint Learning in Innovation Systems in African Agriculture (JOLISAA), a SSA and European Union initiative that brought together several ARD institute in Africa and Europe to generate understanding on how SSA agricultural innovation systems involving smallholder farmers made five policy recommendations (Waters-Bayer *et al.*, 2013). One of them encourages synergies between different knowledge systems:

> *Combining local and external knowledge and resources is necessary to meet the challenges faced by today's and tomorrow's agriculture. It is however, not a simple matter of parachuting in (or transferring) 'scientific' technologies from elsewhere ... Linking the knowledge and ideas of smallholders, small and medium-sized entrepreneurs and local government with external knowledge and ideas leads to an improved capacity and potential for all involved to address ongoing or emerging challenges and opportunities ... Communication and sharing has to be enhanced at different levels, using a range of approaches and methods.*
>
> (ibid.: *pp. 2-3)*

In the above sense, DWR is one of the promising approaches while CL workshops can be seen as the associated method.

Interventionist research methodologies such as DWR are also becoming increasingly necessary to increase the relevance of universities. For example, Kibwika's (2006) study suggests that the relevance of African universities can be achieved through developing their innovation competencies. He notes:

> *The complexity of development problems facing a country like Uganda now requires universities to develop a different form of learning. Universities themselves need to learn to influence change in a complex environment. Most critical is the issue of their competence to provide training, research and outreach services that appropriately address real-life problems ... African universities are increasingly expected to take on societal and developmental functions, in addition to their academic functions.*
>
> *(Kibwika, 2006: p. 165)*

4.2 Designing DWR interventions

A research design deals with the logical dimensions of the study that should be taken into account at the planning (design) stage. This includes research objectives, research sites, research orientation, data generation and analysis methods, and use of research results. DWR research interventions tend to be concerned with:

- research objectives that have an emancipatory interest and seek to improve the situations of research participants;
- research sites that are found in real life or natural settings;
- theoretical concepts associated with improving people's capacities to manage and improve their practices;
- social theories that enable collaborative planning, learning and action;
- research results that stimulate the empowerment of the marginalised, transform practices and communities towards more just and sustainable states; and
- a researcher who deals with dialectical matters such as theory and practice, structure and agency, the past and the future, the local and the national, knowledge that is there and that is yet to be there.

4.3 DWR, power and transformation

Some scholars have argued that DWR has been utilised without paying attention to power relationships among the multiple interacting stakeholders (Kontinen, 2013; Lave, 2012). They note that this normally happens when the focus is limited to technically improving work practice. When this happens, the methodology distances itself from its philosophical roots of transformation and agency development. Transformation here refers to fundamental social and economic changes that empower the poor, marginalised and less resourced members of society. In this regard, Kontinen (2007) recommends the inclusion different voices in the multi-voiced activity systems along the expansive learning process. There are at least four ways that developmental work researchers can use to tackle issues of power and social transformation. These are discussed below.

4.3.1 Negotiate power from the beginning of the intervention

The questions of power have to be partly negotiated before the actual research when the mandate for research is acquired. When mirror data is generated from the concerned actors, it should come from different layers of the organisation such as junior and senior staff and from different interacting activity systems such as farmer organisations, research and extension organisations and policy makers. The conflicts, disturbances and problematic situations that are raised tend to capture matters of power relations. This point is well illustrated by Lindley (2014) whose study concludes that the process of generating mirror data addresses power imbalances by affording the less powerful a voice during ethnographic research and by focusing on the structural contradictions that give rise to problematic situations and not on who causes them. He says:

> *The process of presenting the emerging tensions and contradictions as mirror data during the first intervention workshop has contributed to the strengthening of the democratisation of decision-making in the group. This contributed to the increased participation by Black participants who were much quieter that their White colleagues. While this could be seen as perpetuating racial differentiation, I found it necessary to name people by race as the racial patterns of group dominance and exclusion were so obvious in the data. This is particularly interesting in a South African context where historical forms of dominance continue, despite democratic intentions ... Black participants were predominately very vocal during the interviews and came up with valuable insights, but relatively quiet during the workshop despite my conscious efforts as the facilitator to draw them into conversations. In most situations this would*

> *have excluded their ideas from discussions. However, with the guidance from expansive learning cycle, and through the process of anonymously mirroring the tensions and possible solutions generated from the interviews back to all participants during the workshop, the ideas of Black participants were noticed and taken up in the discussions ... The consequence was that the voices of Black participants were made visible and this allowed for their ongoing inclusion and participation.*
>
> *(Lindley, 2014: pp. 236-237)*

In addition, the resolution of these problematic situations should address power issues where they are present. In this regard Virkkunen (25 September 2014, personal communication) notes:

> *When analysed concretely, a genuine resolution of a contradiction in a productive activity is often a progressive option to attempts to master the contradiction beyond one-sided use of power.*

4.3.2 Developing the shared object and model solution

The developmental work researcher should create space for each stakeholder group to co-create the shared object and to model the solution. This may mean creating moments of group work where interests are articulated and the goals of each group of stakeholders are acknowledged. Each stakeholder group should be given the opportunity to contribute to the development of a model solution. The developmental work researcher may need to spend more time assisting the traditionally marginalised stakeholders to articulate their intended outcomes and model solution.

4.3.3 Pay more attention to issues of structure and agency

Within CHAT-informed DWR, structural issues can be identified through paying attention to causal mechanisms and explanations. Such matters are identified during ethnographic research, which links local issues to the broader context, including the historical. Later they can be identified through identifying rule-producing activity systems and the challenges they may be posing. The third area where structural matters can be identified is in the identification of underlying causes. In this regard, Seppänen (2002) underlines the importance of seeking out the invisible aspects that underlie problematic situations (Figure 4.1). Other researchers have augmented the methodology with frameworks such as critical realism that focus on structure, agency and transformation (Lindley & Lotz-Sisitka, 2014; Mukute, 2010a).

Level of focus	Problems	Solutions
Invisible system activity	Developmental process to identify system contradictions 2	Designing new forms of activity (e.g. new rules, new tools) 3
Visible individual actions, events	1 Identifying the obvious problems	4 Implementing the obvious (or new) solution

Figure 4.1. Problem solving process that includes structural matters (Seppänen, 2002).

4.3.4 Make the less powerful the central activity system

One way that has been used to address different power relations between multistakeholders has been to work with the less powerful first. Later, the less powerful engage with others, either on their own or through the facilitation of a developmental work researcher. Among other things, the developmental work researcher should help the less powerful to come up with critical and coherent conceptions of their situations (Gramsci, 1971). They do this by articulating their issues better, linking them to history and context. For example, Mukute (2013) worked with 32 newly established organic farmer associations that could be considered less powerful in relation to service providers, local leaders, transporters and buyers. He worked with each organic association as a central activity system and helped them to frame their intentions, identify obstacles and develop plans of learning and action. Some of the groups successfully implemented their reconceptualised agricultural activity systems by effectively engaging with the more powerful local leaders and agro-dealers.

4.4 Data generation through activity systems analyses

Successful DWR will partly depend on good data generation and analysis. In DWR, it is important to make a clear distinction between activity system analysis as a way of data generation; and the subsequent stage of data analysis proper. In DWR data generation is guided by an appreciative inquiry, which sensitises the data revealing, among other things:

- visions of valued and possible futures (Textbox 4.1);
- existing and unexplored potentials and innovations;
- lived values and traditions;
- what gives life to a community or practice when it is thriving;
- metaphors that are in circulation;
- a sense of optimism about people's capacity to influence the future; and
- expansion of choices, tools and capacities to act on the world responsibly.

The value of an appreciative inquiry approach is that local and other existing innovations get recognised and utilised. So do the resources and resourcefulness of the local actors and other relevant stakeholders.

There are three main approaches to data generation in DWR and they are presented in the form of analyses (Holzkamp, 1983 cited in Engeström & Engeström, 1986). These are object-historical analysis, theory-historical analysis and actual-empirical analysis. These approaches to data generation underscore the fact that human activity is not only object-oriented and mediated by conceptual tools, but also historical.

The object-historical analysis and the theory historical analysis are jointly called historical-genetic analyses. It is important to note that some specific methods can be used in each of these approaches. The methods include and are not limited to,

Textbox 4.1. The concept of vision of the future in DWR.

A vision of the future in DWR is the model of a qualitatively new system, rather than a set of abstract desiderata. It is based on more materialistic understanding of DWR, which is based on an understanding of the historical development of the activity system, activity or practice and its realistic possibilities of future development. The idea of future possibility is referred to as the Zone of Proximal Development, which is discussed in preceding chapters. Unless there is a slippage in cultural evolution or a contraction in the activity system, the new activity system, activity or practice should be qualitatively better than the previous one. For example, sustainable agriculture is intended to be qualitatively different from conventional agriculture through more responsible use of scarce and diminishing resources such as oil; wider diversity of crops and livestock that recognises plural ways of knowing and contributes to nutrition security. But a vision in the dominant paradigm of change and transformation is based on idealistic thinking in which participants first imagine and describe a good future based on abstract desiderata and then develop a road map towards it.

document analysis, historical time-lines, interviews and focus group discussions and CL workshops. Below are descriptions of three forms of analysis that sensitise data generation in DWR. What is important to note about them is that they result in the organisation of data into *existing* categories or elements of an activity system. Who conducts the analyses during this data generation stage? The DWR researcher and the practitioners or research participants partly collaboratively analyse the data. Practitioners collect part of the data and are also involved in analysing the data that is collected by the researcher, as well as the data that they collect.

The historical analysis produces an understanding of the historical layers of the activity and especially a hypothesis of the central inner contradictions in the activity system.

4.4.1 Object-historical analysis

The object-historical analysis focuses on the previous developmental phases by examining the changing object in the history of the activity or practice (Engeström & Engeström, 1986; Mäkitalo, 2005). An object-historical analysis examines these changes and seeks out their explanations and establishes how they may have influenced other parts of the activity system in order to identify contradictions. For example, the object of home cleaning when housing standards, hygiene and medical care was low in family homes was to manage diseases (e.g. tuberculosis) and parasites. But the object changed to visible cleanliness when medical care hygiene and housing improved (Engeström & Engeström, 1986). Later the object of cleaning included factories and offices. These changes in the object had implications on the tools, rules, community, subjects and division of labour (*ibid.*).

4.4.2 Theory-historical analysis

This involves studying the development and influence of the conceptual tools (theories, methods, models, artifacts) – significant developmental ideas that have been used in the history of the practice or activity for its mastery (*ibid.*). Put differently, this refers to the dominant thought forms that accompanied the historical evolution of the activity. The weight that is given to this kind of analysis depends on the complexity of the practice (Engeström & Engeström, 1986) – hunting may need less theory-historical analysis compared to agriculture. The analysis then culminates in the identification of contradictions in the dominant thought forms and between the dominant thought forms and the practice over time. For example, Mäkitalo's (2005) study of care for the elderly in Finland identified two important theories: one around the ideology of care and the other on the structure of the care services. The ideology of rule care shifted from standardized rule-oriented care in which the client was passive and on the receiving end (which had its origins in caring for the poor); to customized negotiated care, which recognised the individuality and autonomy of each client as well as the activation and rehabilitation of the elderly (*ibid.*).

4.4.3 Actual-empirical analysis

Actual-empirical analysis focuses on the analysis of the present form(s) of the activity system(s), which is conducted under the assumption that the current activity system carries a mix of past and current aspects of the practice under review. It is interested in establishing how practitioners presently do their work and think about it (Engeström & Engeström, 1986). This results in the production of categories of descriptions of work orientations at the levels of activity, action and operations. The categories are constructed in relation to the object-historical and theoretical-historical analyses of an activity.

4.4.4 Relationships between the above forms of analysis

The kinds of data generated through the above methods produce mirror data that should be presented back to the practitioners for validation and strengthening during CL workshops prior to the modelling of solutions. The linkages between the forms of analysis conducted at data generation level are:

- the historical analysis produces an understanding of the historical layers of the activity and especially a hypothesis of the central inner contradictions in the activity system;
- the actual empirical analysis tests and elaborates this hypothesis on the level of daily actions and operations; and
- the object-historical and theory-historical analyses are used to create a model of the Zone of Proximal Development (ZPD) of the activity.

Some of the tools of analysis that have emerged from the creation of ZPD of activity in DWR are:

- extent of collectivity and flexibility (Engeström, 1995);
- degree of change from hierarchy and function-based division of labour to expanded objects of work and responsibilities based on new principles of delegated authority and technology-supported network coordination (Ala-Laurinaho *et al.*, 2009); and
- degree of incremental change or transformative change on one hand, and extent of focus on concept formation or implementation on innovative ideas (Virkkunen & Newnham, 2013).

4.4.5 An example of activity system analysis during data generation

Table 4.1 shows how agriculture could be viewed and analysed as a practice across centuries, showing the evolving nature of the object and the tools and how these were shaped by challenges, many of which were unintended negative effects of progress. While the traditional, agricultural and green revolution rows illustrate the object-historical and theory-historical analysis, the post-green revolution row illustrates the actual-empirical analysis focusing on the object, tools and problematic situations. When moving from one phase to another, it is worth noting that the tools of the previous stage and work orientations do not disappear when the next phase is reached. This tends to create some of the tensions and contradictions that are experienced in the new activity system.

Table 4.1. Illustrations of activity systems analyses during data generation.

Phase	Object	Some major tools (Barrera, 2011; EcoAgriculture Partners, 2012; Cavanagh, 2014; Lotz-Sisitka, 2015; Moore, 2010)	Crises encountered
Traditional agriculture	• Family consumption and survival	• Domestication and selection of plants and animals • Fire, machetes and knives (to open and fertilize land), hoe and plough • Windmill and draught power (energy) • Fallowing (fertility) • Experiential knowledge, trial and error experimentation	• Labour intensiveness • Growing (sedentary) populations • The emergence of the industrial revolution and urbanisation
Agricultural revolution linked to the Industrial revolution	• Family consumption • Sale of surplus for income and industrial raw material	• Hybrid seed and livestock breeding • Farm implements, and machinery (e.g. seed drill, cotton gin, mechanical reaper) • Steam engine, carbon-based energy and electric motors • Crop rotation, chemical fertilisers and pesticides, antibiotics • Food processing and preservation • Agricultural science and promotion of methods and inputs	• Rapid population growth • Declining land for agriculture • Rural-urban migration of displaced farm workers

>>>

Table 4.1. Continued.

Phase	Object	Some major tools (Barrera, 2011; EcoAgriculture Partners, 2012; Cavanagh, 2014; Lotz-Sisitka, 2015; Moore, 2010)	Crises encountered
Green revolution linked to the post WW11	• Family consumption, • Sale of surplus for income, industrial raw material • Export and agribusiness development	• Intensive combination of hybrid seed, chemical fertilisers, pesticides and irrigation • Biotechnology and genetic engineering (across species) • Intensive capital and machinery use and credit facilities • High carbon-based energy use • Agricultural research and development	• Risks and their asymmetrical distribution (e.g. climate change & variability, global warming and eutrophication) • Food shortages, food safety concerns and high food prices in the middle of plenty • Biological resistance to pesticides and herbicides • Shrinking agro-biodiversity, quantitative and qualitative decline of water and land
Post-green revolution	• Food security safety, sovereignty and nutrition • Expanded agriculture – agri-food chains for income • Ecological sustainability and agricultural biodiversity	• Climate change adaptation and mitigation and renewable energy • Biotechnology, ICTs and nanotechnology • Mix of traditional (family), conventional and agro-ecological agriculture • Precision agriculture and precision technologies along the value chain alongside diversity farming • Whole landscape approach and multi-functional agriculture • Green economy • Transdisciplinary, transformative and transgressive learning • Grassroots and multistakeholder innovation	• Sustainability and productivity • Quality and quantity • Social, economic and ecological justice • Wicked problems, risks and uncertainties

4.5 Data analysis

Data analysis refers to the processing of the data that have been generated and organised using a range of approaches and methods discussed above. Data analysis is linked to the specific purpose of a study, which is captured in both the objectives and the research questions. It may lead to the discarding of some of the data generated and/or to the need for further information. Going back and forth between data generation and analysis is a common practice is qualitative research such as DWR

(Merriam, 1998). The data that has been generated may be inadequate to sustain a hypothesis or to drop it.

4.5.1 Phenomenographic analysis

Phenomenographic analysis involves the construction of categories of conceptions that are created partly *inductively* and partly *abductively* (see the next two forms of analysis in DWR described below) using historical types of the activity based on narratives made by the subjects in a study. Phenomenographic analysis is widely used in DWR although reference to it is seldom made. Some of the researchers who have used the analytical framework are Engeström (1987), Simoila (1994) and Mäkitalo (2005). One of the important values of this form of analysis is the recognition of multi-voicedness in practices. The analysis is done in the context of Phenomenography – a study about how people (subjects) experience, understand or conceive of the world around them. It reveals the variations of people's conception of a given phenomenon by interviewing and recording, transcribing, describing, analysing and understanding of experiences (Marton, 1981). Often the entry point of DWR is an analysis of an activity system or current practice that is based on the practitioner's understanding. Data is collected through open-ended in-depth interviews (Khan, 2014). The focus the study is not the phenomenon, and not the experience of the researcher; but the research participants' conceptualisation of the phenomenon. Understanding phenomenon is seen as a relationship between the subject and the object – not as separate and independent (Larsson & Holmströ, 2007). Phenomenography therefore takes a non-dualistic ontological perspective. The different ways of understanding phenomena are captured in *what* the subject's focus is; and *how* meaning is created. Phenomenographic analysis follows several *iterative* steps, which include the following Sjöström & Dahldren (2002):

- reading the transcribed data to establish ways in which participants experience and understand phenomenon;
- identifying relevant answers through selecting parts or extracts that seem to be relevant and meaningful to the objectives of the study;
- clustering similar answers selected during the above processes into tentative categories and progressively refining them;
- giving a name to each cluster or category of data – a name that captures the essence of the category's contents; and
- showing how the different categories may be related (hierarchically).

This results in empirical generalisations by identifying variations, regularities and connections. Textbox 4.2 gives an example of phenomenographic analysis that was conducted with learners.

Textbox 4.2. Phenomenographic analysis with learners.

A typical example that is often used to illustrate phenomenographic analysis is found in the field of education. Marton *et al.* (1993) found six conceptions of learning, which are: (1) increasing one's knowledge; (2) memorising; (3) applying; (4) understanding; (5) seeing in a different way; and (6) personal change. Their analysis concluded that the first three categories focus on accumulation and constitute the *what* of learning, which is quantitative and reproductive. The last three categories are linked to the qualitative view of learning – that is, *how* learning happens, which is a productive view of learning that is closely linked to innovation and cultural evolution. In a study that sought to explore and expand farmer learning in the context of sustainable agriculture and DWR, Mukute (2010) identifies similar categories that were based on inductive analysis:

- acquiring (increasing) knowledge from scientists and fellow farmers who know more;
- applying knowledge in agricultural practice;
- observing weather, crops and croplands, animals and pastures and making sense of them;
- linking everyday knowledge to scientific knowledge; and
- trial and error and experimentation.

4.5.2 Inductive analysis

Inductive analysis is concerned with making sense of the data generated by clustering data into categories based on the notion of 'letting data speak'. Collected data is analysed to gain an understanding of phenomenon. In fact, the forms of analysis that are done under genetic-historical, actual-empirical and phenomenographic analysis are initially based on inductive analysis, which groups like things together. In the process of doing this, variations within and between 'things' in a practice are also revealed. This process provides the bridge between the two spaces of the 'real world' and the represented world.

4.5.3 Abductive analysis

Abductive analysis is based on the use of theoretical lenses to view and make sense of data. This helps the researcher to interpret, re-describe, reframe, recontextualise and reconceptualise the concrete situation from a particular theoretical or conceptual perspective. For example, in CHAT-informed DWR, data may be organised according to the elements of an activity system: object, tools, subject, rules, community, division of labour and contradictions. It could also be viewed from the perspective of empirical and theoretical generalisations or from the angle of agency, learning and development. An important way of applying the concept of hierarchy in DWR research is based on the relationship between activity, action and operations.

What is important to remember is that these different forms of analysis are closely linked to each other and are applied at the different stages of the process of analysis as one stage has a bearing on the other. For example, abductive analysis tends to be done when the others forms of analysis have been done.

4.6 DWR and the context of appreciation

Many scholars have noted that the expansive learning process begins with a process of *questioning* current practice in order to reveal the need state in the context of criticism. Experiences of working with DWR reveal that there is need to acknowledge for the developmental work researcher to first genuinely learn about the practitioners and their practices. The main purpose of such research is to appreciate the cultural history of the practices that are to be considered for an intervention. This process may also involve understanding the other activity systems that the subjects interact with. Therefore, in DWR, prior to helping practitioners to question their practices for possible improvement, there is in fact, a process of understanding and appreciate it. This may be called the *context of appreciation.*

4.7 DWR and timeframes

It is important to provide enough time for each stages of the DWR process. Experiences in Finland and New Zealand suggest that one needs a minimum of two years to complete the expansive learning cycle (Engeström, 2001; Hill *et al.*, 2007). For example, transitioning to some forms of agricultural practices involves farming across years (Seppänen & Koskimies, 2002). Leaving the field during the initial stages of implementing a solution before it has begun encountering inevitable resistance (tertiary and quaternary contradictions) gives a misleading (over-optimistic) view of the intervention (Mukute & Lotz-Sisitka, 2011). Below are some interesting time-related lessons.

4.7.1 Time to develop mutual trust

Adequate time should be set aside to understand the context in which the research is being conducted. At this stage it is important to spend a reasonably long period of time being with the people using ethnographic research. The time needed tends to be longer if one is new to the social, cultural and ecological environment. This is so partly because there is need to gain both physical and social access. The latter refers to getting on with the research participants and generating the necessary trust to share useful information. Some scholars even argue that an interventionist researcher should develop a sense of place, that is, detailed knowledge of a place, a sense of care and rootedness (Barter, 2007). And such sense of place develops better if someone has lived with a community for a certain period of time. Active interviews and close joint analysis of problematic situations enable development of a good rapport of collaboration between the developmental work researcher and participants (Holstein & Gubrium, 1995).

4.7.2 Time for collective learning, innovation and planning

Adequate time should also be set aside for CL workshops during which shared objects are developed, problematic situations analysed and model solutions developed. Each session should be long enough to allow participants to move along the expansive learning process. At the same time, it should be short enough to enable participants to concentrate. Experience suggests that each session can range between 1.5-3.0 hrs. The sessions tend to be longer if participants have significantly different interests, cultures and ways of articulating themselves, and where translations are necessary. The number of sessions along the expansive learning process varies considerably. It is important to dedicate at least one session to each stage of the expansive learning process.

4.7.3 Time for and timing of implementing and reviewing model solutions

Experiences in Finland and New Zealand suggest that one needs a minimum of two years to complete the expansive learning cycle. This has also been confirmed in southern African countries. If progress is reviewed too soon, one may get a misleading view of the impact of the intervention. The view can be either optimistic or pessimistic. For example, in a case study involving the organic farming and value addition, a review of the implementation of the model solutions 12 months later showed farmer associations were not yet addition value to their produce. However, another review conducted 18 months later revealed that some farmer associations were marketing value-added produce.

4.7.4 Time to pass through distances

Physical, social and ideological distances between research participants determine the pace of progress in learning, innovation and collective agency development. The greater the distance, the more energy is required to overcome resistance. People who have to travel far and for longer periods are likely to need more effort to bring together than those living near one another. The same applies to people who belong to different disciplines or have different social standing, religious or political persuasions. This means that more time, energy and resources should be set aside in interventions that involve multiple stakeholder groups that are not used to working together than for homogenous groups.

4.7.5 Time to change ways of doing things

Transforming agricultural practices so that they become more sustainable takes time for several reasons. There is need for time to master the necessary knowledge and skills among farmers. There is also need for time to build the soil, establish agro-biodiversity produce ecological services beyond the immediate farm. Altieri (1995) identifies four phases of transition to sustainable agriculture: (1) progressive chemical withdrawal; (2) rationalized agro-chemical use through integrated pest management and nutrient management; (3) input substitution; and (4) redesign of the farming system to encourage diversity, synergy, so that the system can sponsor its own soil fertility, natural pest regulation and productivity. Aschmann & Cutforth (2000) suggest that it may take 3 to 6 years to fully implement a sustainable system.

The need for time to combine related actions to improve an activity through addressing its object is highlighted and explained by a CHAT theorist:

> *In activity theory, the distinction between individual goal-directed action and collective object-oriented activity is of central importance. Individual actions may be depicted as focusing on their linear dimension or focusing on their socio-spatial dimensions. In any case, the temporal duration of actions is relatively short. Development takes place in long historical cycles, typical of the evolution of collective activities ... A collective activity is driven by a communal motive. The motive is formed when a collective need meets an object that has the potential to fulfil the need. The motive is thus embedded in the objective of the activity. The object, in turn, is to be understood as a project under construction, moving from the potential 'raw material' to a meaningful shape and to a result or an outcome.*
>
> *(Engeström, 1999: p. 65)*

4.8 Developmental work researcher and participants

Developmental work researchers need to make an effort to make the situational analysis, contradiction and analysis and solution generation process collective and ensure that it is owned by practitioners. At the same time, the developmental work researcher will stimulate the expansive learning process better by minimising or avoiding a strong researcher-practitioner duality. This is achieved partly through ensuring that the practitioners play certain roles during the CL workshop sessions. One of the key roles is that of documenting the analysis. This can be achieved through participants capturing issues and ideas from CL discussions on flip charts. Below is an illustration of how CL research participants used the flipchart to highlight important points arising from discussions. For example, in a CL workshop that was conducted at St Margaret School in 2009, the teachers who promoted permaculture developed the diagram shown in Figure 4.2 as their activity system. The main change that was later made to it concerned the subjects, who were changed to teachers only.

Figure 4.3 summarises how research participants in the Isidore Organic Agriculture case study captured the desire for 'quick money' as one of the explanations behind agricultural challenges being encountered in their country and beyond. This was

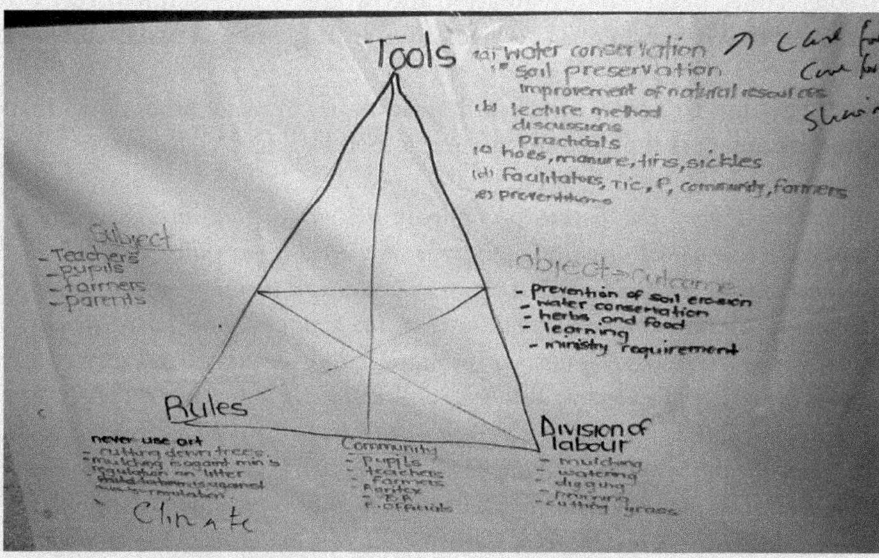

Figure 4.2. Teachers' construction of the St. Margaret Permaculture activity system.

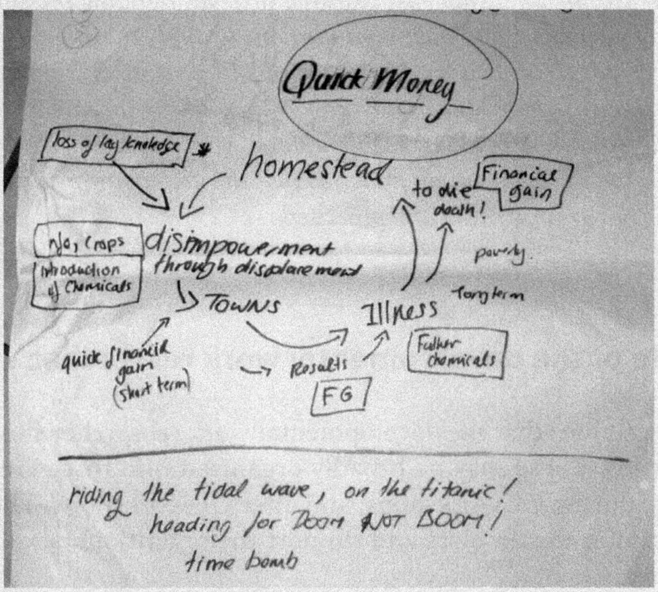

Figure 4.3. Isidore Organic community analysis of agriculture challenges.

later traced to be a contradiction between short-term economic gains, social justice and long-term ecological sustainability. This is a typical sustainable development contradiction.

The Noah's Ark solution was a Durban Organic Forum that comprised an inner circle of: organic farmers, trainers and mentors, retailers, organic certifiers and local government. The outer circle comprised key stakeholder groups that constituted the community as we as the providers of tool and rules.

4.9 DWR and theoretical generalisation

The idea of theoretical generalisation is becoming more widely used in agricultural innovation thinking. However, the terminology that is used is different – scaling up or institutionalisation. It is therefore essential to connect with these more commonly used concepts. The linkage is illustrated in the following statement:

> *Vertically scaling up farmer-led research involves building the capacity of different stakeholders and their organisations to apply the approach as part of their regular work. It is a complex process that requires stimulating and enabling change in individuals and, through them, in organisations. It requires broad multistakeholder alliances pursuing an agreed theory of change ... care must be taken to embed the approach so that the underlying principles of the ... approaches are embraced and practised.*
>
> (*Waters-Bayer* et al., *2015: p. 5*)

4.10 The role of the developmental work researcher

There are several roles that the developmental work researcher should play along the DWR journey. These may be broadly organised into three areas: before the CL workshops; during CL workshops; and after CL workshops. The basic idea of a CL intervention is to stimulate and support the practitioners to take expansive learning actions.

4.10.1 Before CL workshops

At this stage the developmental work researcher should plan the intervention and conduct research. The tasks can be summarised as follows:

- Conduct document analysis and literature review concerning the people and activity systems selected for an intervention in order appreciate their history, contexts and relevant practices.
- Visit and engage research participants to get more information about the activity systems of interest. Analyse the secondary and primary data and produce 'mirror data' for presentation to research participants.
- Prepare workshop materials on the methodology and plan for the documentation of the CL workshop.
- Choose a suitable venue for the workshop – one where research participants will find conducive for learning and innovation.

4.10.2 During CL workshops

Some of the main tasks of the developmental work researcher during CL workshops are:

- Facilitating the creation of an atmosphere that encourages different ways of knowing to be valued, appreciated and utilised. There is need to pay particular attention of marginalized knowledges and ways of knowing.

- Presenting mirror data to highlight some of the main challenges faced in activities under review.
- Developing the participants' capacities to work with the methodology.
- Facilitating participants dialogue around questioning, analysing, modelling and examining solutions.
- Probing participants so that they articulate problematic situations, innovations, objects and intended outcomes.
- Clarifying and suggesting based on what participants say. This is based on the developmental work researcher's broader systems perspective, and understanding of the theoretical concepts associated with the methodology.
- Documenting workshop deliberations, which could include video-recording and transcribing for subsequent analysis.
- Facilitating the development of action plans concerning the implementation of the innovative solutions.
- Reviewing the implementation of the innovations, associated learning and agency development. This should be accompanied by identifying emerging contradictions and modelling new solutions to resolve them and consolidate the new practice.

4.10.3 Between and beyond CL workshops

Some of the most important tasks of the developmental work researcher take place outside the CL workshops. These include:

- Studying video-recorded materials between sessions in order to analyse proceedings and identify important information. Such information could be in the form of breakthroughs, contradictions or agentive or reflective talk. It is often presented in the next session to enrich discussions.
- Reminding participants of their commitments in order to prepare them for subsequent processes. This is particularly important during the process of implementation ahead of review or feedback workshops.
- Reflecting on field experiences and documenting them. One of the reports should capture the learning, innovation and agency development process. It should ideally be shared with research participants for the input. The other report can be of a more theoretical nature for academics, addressing research questions.

4.10.4 Learning, innovation and agency development take place beyond CL workshops

Research studies that used DWR in agriculture, medicine and NRM settings have shown that the bulk of the expansive learning may actually take place outside the workshop when the model solution is being implemented and adjusted and when agency is exercised on policies and institutions. For example, in one case study where the model solution was to establish a district demonstration plots to show policy makers and implementers the potential of a sustainable agriculture practice, research participants met the district leaders more than once to lobby for allocation of a plot. They not only got the plot but also received additional resources and material support to make it work from a local authority that had historically undermined the practice.

4.10.5 Expansive learning transcends top-down and bottom up learning and development

The importance of the faithful transmission of culturally accumulated knowledge as a condition for avoiding slippage in cultural evolution shows that top-down learning is necessary for development. Such knowledge comes from both specialists and lay people (practitioners). The experiences and idiosyncratic knowledge of practitioners is critical for the continued improvement of their activity systems and associated practices – highlighting the need for bottom-up approaches. But what makes DWR special is that it combines these two dialectically related approaches and uses them as the bases of modelling solutions in and between CL workshops.

4.10.6 CL workshops work best when there is longstanding collaboration among practitioners and researcher

Successful DWR and CL workshops intervention require longstanding cooperation, collaboration and communication between the developmental work researchers, the practitioners in the activity systems under review and the associated content specialists. This long-term interaction helps in the development of knowledge about each other, mutual trust and rapport. At the same time, it allows for the building of a series of interconnected smaller cycles of learning that accumulate and constitute a complete expansive learning cycle. This is achieved through overcoming secondary, tertiary and quaternary contradictions, which enables boundary crossing by subjects of the interacting activity systems: practitioners, 'substance' specialists and process specialists. The CL workshop serves as the confluence, the meeting place for mutual enrichment of the tributaries of the different cognitions into a unified river full of constructive energy and life. This insight is illustrated below (Figure 4.4). The confluence, the CL workshop is a place of dialogue – a dialogue that avoids certain attitudes and upholds others as Hope & Timmel (1995: p. 26) drawing on Freire (1970) note:

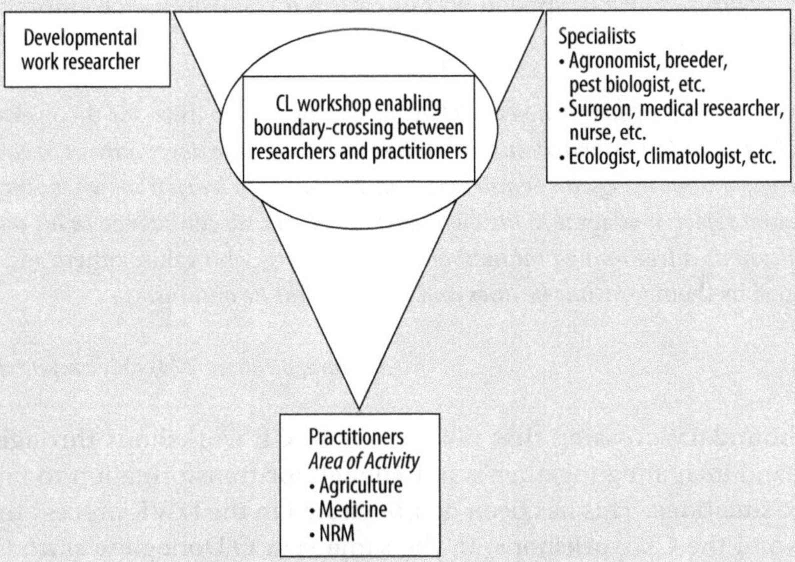

Figure 4.4. A change laboratory as an instrument for practice-inspired multidisciplinary research.

> *How can I enter into dialogue if I always imagine that others are ignorant, and never become conscious of my own ignorance? How can I enter into dialogue if I see myself as a person apart from others – if I see them only as 'its' not as another I? How can I enter into dialogue if I consider myself as a member of the 'in-group', the owners of truth and knowledge ...?* At the point of encounter, there are neither utterly ignorant people, nor perfectly wise people. *There are only those who attempt together to learn more than they know now. [My emphasis]*

It helps if the springs that give rise to the tributaries come from a diversity of cultural landscapes because no single cultural paradigm can adequately address today's complex challenges. Rukuni (2009) identifies the following major cultural paradigms:

- *Western:* 'I think therefore I am', which has enabled its scientific advancement and led the Americans to believe in themselves – 'I am because I can';
- *Eastern:* 'I am because I improve', which has enabled them to improve technologies, organisations and goods; and
- *African:* 'I am because we are', which values relationships between people, including family and community cohesion.

It appears Nyerere's (1968) speech on Education for Self-Reliance embraces these three cultural paradigms:

> *It must produce good farmers ... They have to be able to* think for themselves *... The education provided must therefore encourage the development in each citizen, of three things: an enquiring mind, an ability to learn from what others do, and reject or* adapt *it to his own needs; and a basic* confidence *in his own position as a free and equal member of the society,* who values others and is valued by them *for what he does and not for what he obtains.*
>
> *(Nyerere, 1968: p. 274) [My emphasis]*

But the boundary-crossing that takes place in CL workshops through being, knowing and imagining together is not enough for transformation to take place in real life situations. This has been demonstrated in the DWR process that takes place beyond the CL workshops. In the same vein O'Donoghue *et al.* (2013: p. 124) conclude their book on African Heritage Knowledge in the Context of Social Learning by reminding us that 'Knowing and imagining better things together is not enough unless we can bring about change'. This refers to human agency that is built and exercised along the journey from the abstract to the concrete, which is central in DWR.

4.11 Critical success factors

Several factors enable DWR processes cause learning, innovation, agency, development and transformation. Some of these are:

- longstanding relationships between the DWR researcher, specialists in the field of the activity or practice and practitioners;
- adequate time, financial, material human resources available for the entire expansive learning process;
- an enabling operating environment in the form of policies, structures and systems. These should enable dialogue, engagement, experimentation while at the same time embracing transformative change; and
- the ability to engage with other activity systems and multistakeholders in a manner that enables innovative solutions to be integrated in broader societal practices.

4.12 Conclusions

4.12.1 Concluding the chapter

This chapter sought to provide some insights on conducting DWR in agriculture and natural resources management contexts. It is based on the hands-on experience of conducting such research in studies that focused on improving agricultural practices and enterprise development. The lessons learnt may help the developmental work researcher to be alert to some of the pitfalls that are likely to be encountered in the field. The resource book has explained why it is necessary for actors in the agriculture and NRM sectors to consider use of DWR in their innovation processes. It has also discussed the thinking behind DWR, focusing on the main theoretical concepts that inform it. This was done to provide the theoretical foundation of the methodology, which enables its better and more flexible use. The subsequent explanation of the methodology and the principles that guide it served to provide guidelines on how to work with DWR. The concluding chapter highlighted some of the lessons and insights generated from working with the methodology in the agriculture and NRM sectors. Hopefully, the reader finds the methodology useful for conducting rigorous research, concept development and for improving practices, people's lives and conditions.

In concluding this chapter I am reminded of the introductory remarks in *The future for the excluded – Job creation and income generation by the poor* (Carmen & Sabrado, 2000). The remarks challenge development interventions that stop at stimulating consciousness and knowledge production. Drawing on Leont'ev (1978) they underline the importance of the object of an activity – the practitioners' needs – as the basis for an intervention. But more importantly, they underscore the need for supporting the practitioners in positively transforming their livelihoods, which is the hallmark of interventionist research. This is an important motive of DWR research in the context of agriculture and natural resources management. The other motive is academic.

4.12.2 Concluding the book

It appears fitting to conclude the whole book with some words on DWR's next ZPD, which may be described as transgressive learning in the context of runaway objects that are difficult to define in today's risk society. The current thrust of DWR is and has been on transformative learning associated with Bateson's Learning Level II, which is characterised by *complex problems* that:

- 'resist' solving;
- are unclear and partly familiar because they have interconnected elements and dimensions but can be understood with time;
- have hidden root causes; and
- require non-linear solutions that take time to work out (Gibson, 2013; Wals, 2015).

Transformative learning enables the collective generation of solutions by multiple stakeholders and can therefore be solution-oriented. It built on or grew out of learning processes that were based on acquiring professional competencies to address *simple problems* or that are characterised by a clear problem with a clear solution, predictable, straightforward and obvious outcomes, which are typical of Learning Level I. Administering proven vaccination against a livestock disease is an example of such a problem.

But DWR's next concern is located in Learning Level III, which is to be found in the emerging fourth generation CHAT that was not discussed in this book partly because it is still forming and has not yet crystallised. *Wicked or highly complex problems* (some of which are addressed through transformative learning described in the book) that have runaway objects demand another form of learning – transgressive learning. It is associated with resolving wicked problems that are:

- elusive or difficult to define and even more difficult to solve – without a clear 'stopping point' in responding to them;
- involve multiple stakeholders with conflicting perspectives and values;
- resolutions of complex problems generate unexpected consequences over time;
- involve changes in belief, behaviour and/or identity; and
- have no right or wrong (re)solutions only good and bad, and on which judgment can only be made several years after their implementation (Rittel & Webber, 1973; Gibson, 2013).

Dealing with such wicked problems requires *radical* cultural and societal change and an ability to deal with denial of the problem, despair of dealing with the problem and 'grief' at the loss of past behaviours (Preece, 2003 cited in Kronlid & Friman, 2015). Engeström (2008) identifies the need for the *radical transformation* of activity systems in the context of highly complex postmodern society, which is akin to transgressive learning. His starting point is the re-defining of the object – the wicked problem, which he describes as a runaway object. He reminds the reader that, CHAT in which DWR was developed, the object is a concern or a generator and focus of attention, motivation, effort and meaning. He further notes that in this era, the major concerns may be described as 'runaway objects' underlining their elusive and dynamic nature. They are pervasive and they generate opposition and controversy. They also 'often remain dormant, invisible, or unseen for lengthy periods of time, until they burst out into the open in the form of acute crises or breakthroughs' (*ibid.*: p. 4). He goes on to say:

> *With runaway objects, we may ask: Are there objects without an activity? Whose object is global warming for example? Of course runaway objects do not emerge and exist without human activities ... The very concept of global warming would not exist if experts, researchers, politicians and journalists did not articulate the phenomenon. But which activities [and activity systems] take the responsibility for such a huge object as global warming?*
>
> *(Engeström, 2008: p. 4)*

The next task of CHAT and DWR is therefore to work out ways of dealing with wicked problems. In the meantime, it is worth considering use of DWR in dealing with the complex problems that require transformative and transdisciplinary learning involving practitioners, content and process specialists working together.

References

Ala-Laurinaho, A., Schaupp, M., & Kariniemi, A. (2009). The qualitative differences of the effect of technological changes: Case wood procurement process. In: L. Norros, H. Koskinen, S. Salo & P. Savioja, (Eds.), *Designing beyond the product – Understanding activity and user experience in ubiquitous environments. Proceedings of ECCE 2009 European Conference on Cognitive Ergonomics* (pp. 289-293). Helsinki, Finland: VTT.

Altieri, M. (1995). Agroecology: *The science of sustainable agriculture* (2nd ed.). London, UK: ITDG.

Aschmann, S., & Cutforth, L. (2000). Making the transition to sustainable agriculture. *Ecological Sciences Technical Note 2, Agronomy.*

Barrera, A. (2011). *New realities, new paradigms: the new agricultural revolution.* San-Jose, Costa Rica: International Institute for Cooperation in Agriculture (IICA).

Barter, B. (2007). Rural education: Learning to be rural teachers (pp. 59-64). Paper 245. *Conference Proceedings on Researching work and learning. 5th International Conference on Researching Work and Learning (RWL5).* December 2-5, 2007. Cape Town, South Africa: University of Cape Town, Human Sciences Research Council, University of the Western Cape & SAQA.

Carmen, R., & Sabrado, M., (Eds.) (2000). *The future for the excluded – job creation and income generation by the poor.* London, UK: Zed Books.

Cavanagh, C.J. (2014). Biopolitics, environmental change, and development studies. *Forum for Development Studies, 41 (2),* 273-294.

Danermark, B., Ekstrom, M., Jakobsen, L., & Karlsson, J.C. (2002). *Explaining Society: Critical Realism in Social Sciences.* London, UK: Routledge.

EcoAgriculture Partners (2012). *Landscapes for people food and nature: The vision, the evidence and next steps – an international initiative for dialogue, learning and action.* Washington, DC, USA: EcoAgriculture Partners.

Engeström, Y. (1987). *Learning by expanding: An activity theoretical approach to developmental research.* Helsinki, Finland: Orienta-Konsultit.

Engeström, Y. (1995). Objects, contradictions and collaboration in medical cognition: an activity-theoretical perspective. *Artificial Intelligence in Medicine, 7,* 395-412.

Engeström, Y. (1999). Expansive visibilisation of work: An activity-theoretical perspective. *Computer Supported Cooperative Work, 8,* 63-93.

Engeström, Y. (2001). Expansive learning at work: Towards an activity theoretical re-conceptualisation. *Journal of Education and Work, 14 (1),* 133-156.

Engeström, Y. (2008). *From design experiments to formative interventions.* Paper Presented at the ISCAR Conference. September 9-13, 2008. San Diego, CA, USA: University of California.

Engeström, Y., & Engeström, R. (1986). Developmental work research: The approach and the application of cleaning work. *Nordisk Pedagogik, 6,* 2-15.

Freire, P. (1970). *Pedagogy of the oppressed.* New York, NY, USA: Herder and Herder.

Gibson, R. (2013). *Problems: simple, complex, wicked.* Available at: http://tinyurl.com/qfwxnn7.

Gramsci, A. (1971). *Selections from the prison notebooks.* Edited and translated by Hoare, Q. & Smith, G.N. London, UK: Lawrence and Wishart.

Hill, R., Capper, P., Wilson, K., Whatman, R., & Wong, K. (2007). Workplace learning in the New Zealand apple industry network: A new co-design method for government 'practice making'. *Journal for Workplace Learning, 19 (6),* 359-376.

Holstein, J.A., & Gubrium, J.F. (1995). The active interview. Qualitative research methods series, volume 37. London, UK: Sage.

Hope A., & Timmel, S. (1995). *Training for transformation: A handbook for community workers – Book II.* Gweru, Zimbabwe: Mambo Press.

Khan, S.H. (2014). Phenomenography: A qualitative research methodology in Bangladesh. *International Journal on New Trends in Education and Their Implications, 5 (2),* 34-43.

Kibwika, P. (2006). *Learning to make change: Developing innovation competence for recreating the African university of the 21st century.* Wageningen, the Netherlands: Wageningen Academic Publishers.

Kontinen, T. (2007). *Learning changes of NGOs in development: Cooperation of Finnish NGOs in Morogoro, Tanzania.* Helsinki, Finland: Helsinki University Press.

Kontinen, T. (2013). A Gramscian perspective on developmental work research: Contradictions, power and the role of researchers reconsidered. *Outlines – Critical Practice Studies, 14(2),* 106-129.

Kronlid, D., & Friman, E. (2015). *Think piece on transformative, transgressive and transdisciplinary learning for climate change adaptation.* Paper presented at the Research Design Workshop Research Workshop. January 18-21, 2015. Grahamstown, South Africa: Rhodes University

Larsson, J., & Holmströ, I. (2007). Phenomenographic or phenomenological analysis: Does it matter? Examples from a study on anaesthesiologists' work. *International Journal of Qualitative Studies on Health and Wellbeing, 2,* 55-64.

Lave, J. (2012). Changing practice. *Mind, Culture and Activity, 19,* 156-171.

Leont'ev, A.N. (1978). *Activity, consciousness and personality.* Eaglewood Cliffs, NJ, USA: Prentice-Hall.

Lindley, D. (2014). *Can expansive (social) learning processes strengthen organisational learning for improved wetland management in a plantation forestry company, and if so how? A case study of Mondi.* Unpublished PhD thesis. Grahamstown, South Africa: Rhodes University.

Lindley, D., & Lotz-Sisitka, H. (2014). *Expansive social learning, morphogenesis and reflexive action in an organisation responding to wetland degradation.* Paper presented at the 30th EGOS (European Group for Organisational Studies) Colloquium. July 2-5, 2014. Rotterdam, the Netherlands: Rotterdam School of Management, Erasmus University.

Lotz-Sisitka, H. (2015). *Tribute and memorial lecture: a tribute to Professor Ben Parker.* Memorial Lecture made in Pretoria, South Africa on 22 January 2015.

Mäkitalo, J. (2005). *Work-related well-being in the transformation of nursing home work.* Oulu, Finland: University of Oulu.

Marton, F. (1981). Phenomenography: describing conceptions of the world around us. *Instructional Science, 10(2),* 177-200.

Marton, F., Dall' Alba, G., & Beaty, E. (1993).Conceptions of learning. *International Journal of Educational Research 19 (3),* 277-300.

Moore, J.W. (2010). The end of the road? Agricultural revolutions in the capitalist world-ecology, 1450-2010. *Journal of Agrarian Change, 10 (3),* 389-413.

Merriam, S.B. (1998). *Qualitative research and case study applications in education.* San Francisco, CA, USA: Jossey-Bass Publishers.

Mukute, M. (2010). *Exploring and expanding farmer learning in sustainable agriculture workplaces.* Unpublished PhD thesis. Grahamstown, South Africa: Rhodes University.

Mukute, M. (2013). Bridging and enriching top-down and participatory learning: the case of smallholder, organic conservation farmers in Zimbabwe. *Southern African Journal of Environmental Education, 29,* 74-92.

Mukute, M., & Lotz-Sisitka, H. (2011). *Reflections on the use of Change Laboratory workshops in sustainable agriculture workplace contexts in Africa.* International Society for Cultural and Activity Research (ISCAR) Congress. September 5-10, 2011. Rome, Italy: University of Rome.

Nyerere, J. (1968). *Education for Self-Reliance in Nyerere 1968 Freedom and Socialism, Uhuru naUjamaa.* Dar es Salaam, Tanzania: Oxford University Press.

O'Donoghue, R., Shava S., & Zazu C., (Eds.) (2014). *African heritage knowledge in the context of social learning: Learning contributions of the Regional Centres of Expertise on education for sustainable development.* Yokohama, Japan: United Nations University, Institute of Advanced Studies.

Rittel, H.W.J., & Webber, M.M. (1973). Dilemmas in general theory of planning. *Policy Sciences, 4,* 155-169.

Rukuni, M. (2009). *Leading Africa.* Pretoria, South Africa: *Mandala Publishers.*

Seppänen, L. (2000). Activity theoretical view on crop rotation planning in organic vegetable farming. (pp. 283-292) In: *Proceedings of the Fourth European Symposium on European Farming and Rural Systems Research and Extension*, in Volos, Greece, April 3 to 7, 2000.

Seppänen, L. (2002). Creating tools for farmers' learning: An application of developmental work research. *Agricultural Systems, 73(1)*, 129-145.

Seppänen, L., & Koskimies, H. (2002). Farming across the years: Temporal and spatial dimensions in learning organic farming (pp. 528-537). *Proceedings on IFSA European Symposium on Farming and Rural Systems Research and Extension: Local identities and globalisation*. April 8-11, 2002. Florence, Italy: IFSA.

Simoila, R. (1994). *The development, contradictions and work orientations of the work of the public-health nurse*. Sosiaali – ja terveysalan tutkimus – ja kehittämiskeskus. Tutkimuskia 40. Gummerus kirjapaino Oy, Jyväskylä.

Sjöström, B., & Dahldren, L.O. (2002). Applying phenomenography in nursing research. *Journal of Advanced Nursing, 40 (3)*, 399-345.

Virkkunen, J., & Newnham, D.S. (2013). *The change laboratory: A tool for collaborative development of work and education*. Rotterdam, the Netherlands: Sense Publishers.

Wals, A.J.E. (2015). *Think piece on transformative, transgressive and transdisciplinary learning for climate change adaptation*. Paper presented at the Research Design Workshop Research Workshop. January 18-21, 2015. Grahamstown, South Africa: Rhodes University.

Waters-Bayer, A., Kristjanson, P., Wettasinha, C., Van Veldhuizen, L., Quiroga, G., Swaans, K., & Douthwaite, B. (2015). Exploring the impact of farmer-led research supported by civil society organisations. *Agriculture and Food Security, 4 (4)*, 1-7.

Waters-Bayer, A., Triomphe, B., & Oudwater, N. (2013). Building on local dynamics: 5 policy recommendations for enhancing innovation by African smallholder farmers. *Joint Learning in Innovation Systems in African Agriculture (JOLISAA) Policy Brief July 201 3*, 1-6.

www.ingramcontent.com/pod-product-compliance
Lightning Source LLC
Chambersburg PA
CBHW061215240926
51846CB00006B/1239